Learning ROS for Robotics Programming

A practical, instructive, and comprehensive guide to introduce yourself to ROS, the top-notch, leading robotics framework

Aaron Martinez

Enrique Fernández

[PACKT] PUBLISHING open source community experience distilled

BIRMINGHAM - MUMBAI

Learning ROS for Robotics Programming

First published: September 2013

Production Reference: 1190913

Published by Packt Publishing Ltd.
Livery Place
35 Livery Street
Birmingham B3 2PB, UK.

ISBN 978-1-78216-144-8

www.packtpub.com

Cover Image by Duraid Fatouhi (duraidfatouhi@yahoo.com)

Credits

Authors
Aaron Martinez
Enrique Fernández

Reviewers
Luis Sánchez Crespo
Matthieu Keller
Damian Melniczuk

Acquisition Editors
Kartikey Pandey
Rubal Kaur

Lead Technical Editor
Susmita Panda

Technical Editors
Jalasha D'costa
Amit Ramadas

Project Coordinator
Abhijit Suvarna

Proofreader
Joanna McMahon

Copy Editors
Alfida Paiva
Mradula Hegde
Gladson Monteiro
Sayanee Mukherjee
Adithi Shetty

Indexers
Hemangini Bari
Rekha Nair

Graphics
Ronak Dhruv

Production Coordinator
Manu Joseph

Cover Work
Manu Joseph

About the Authors

Aaron Martinez is a computer engineer, entrepreneur, and expert in digital fabrication. He did his Master's thesis in 2010 at the IUCTC (Instituto Universitario de Ciencias y Tecnologias Ciberneticas) in the University of Las Palmas de Gran Canaria. He prepared his Master's thesis in the field of telepresence using immersive devices and robotic platforms. After completing his academic career, he attended an internship program at The Institute for Robotics in the Johannes Kepler University in Linz, Austria. During his internship program, he worked as part of a development team of a mobile platform using ROS and the navigation stack. After that, he was involved in some projects related to robotics, one of them is the AVORA project in the University of Las Palmas de Gran Canaria. In this project, he worked on the creation of an AUV (Autonomous Underwater Vehicle) to participate in the Student Autonomous Underwater Challenge-Europe (SAUC-E) in Italy. In 2012, he was responsible for manufacturing this project; in 2013, he helped to adapt the navigation stack and other algorithms from ROS to the robotic platform.

Recently, Aaron created his own company called Biomecan. This company works with projects related to robotics, manufacturing of prototypes, and engineering tissue. The company manufactures devices for other companies and research and development institutes. For the past two years, he has been working on engineering tissue projects, creating a new device to help researchers of cell culture.

Aaron has experience in many fields such as programming, robotics, mechatronics, and digital fabrication, many devices such as Arduino, BeagleBone, Servers, and LIDAR, servomotors, and robotic platforms such as Wifibot, Nao Aldebaran, and Pioneer P3AT.

I would like to thank my girlfriend who has supported me while writing this book and gave me motivation to continue growing professionally. I also want to thank Donato Monopoli, Head of Biomedical Engineering Department at ITC (Canary-Islands Institute of Technology), and all the staff there. Thanks for teaching me all I know about digital fabrication, machinery, and engineering tissue. I spent the best years of my life in your workshop.

Thanks to my colleagues in the university, especially Alexis Quesada, who gave me the opportunity to create my first robot in my Master's thesis. I have learned a lot about robotics working with them.

Finally, thanks to my family and friends for their help and support.

Enrique Fernández is a computer engineer and roboticist. He did his Master's Thesis in 2009 at the University Institute of Intelligent Systems and Computational Engineering in the University of Las Palmas de Gran Canaria. There he has been working on his Ph.D for the last four years; he is expected to become a Doctor in Computer Science by September 2013. His Ph.D addresses the problem of Path Planning for Autonomous Underwater Gliders, but he has also worked on other robotic projects. He participated in the Student Autonomous Underwater Challenge-Europe (SAUC-E) in 2012, and collaborated for the 2013 edition. In 2012, he was awarded a prize for the development of an underwater pan-tilt vision system.

Now, Enrique is working for Pal-Robotics as a SLAM engineer. He completed his internship in 2012 at the Center of Underwater Robotics Research in the University of Girona, where he developed SLAM and INS modules for the Autonomous Underwater Vehicles of the research group using ROS. He joined Pal-Robotics in June 2013, where he is working with REEM robots using the ROS software intensively and developing new navigation algorithms for wheeled and biped humanoid robots, such as the REEM-H3 and REEM-C.

During his Ph.D, Enrique has published several conference papers and publications. Two of these were sent to the International Conference of Robotics and Automation (ICRA) in 2011. He is the co-author of some chapters of this book, and his Master's Thesis was about the FastSLAM algorithm for indoor robots using a SICK laser scanner and the odometry of a Pioneer differential platform. He also has experience with electronics and embedded systems, such as PC104 and Arduino. His background covers SLAM, Computer Vision, Path Planning, Optimization, and Robotics and Artificial Intelligence in general.

I would like to thank my colleagues in the AVORA team, which participated in the SAUC-E competition, for their strong collaboration and all the things we learned. I also want to thank the members of my research group at the University Institute of Intelligent Systems and Computational Engineering and the people of the Center of Underwater Robotics Research in Girona. During that time, I expended some of the most productive days of my life; I have learned a lot about robotics and had the chance to learn player/stage/Gazebo and start with ROS. Also, thanks to my colleagues in Pal-Robotics, who have received me with open arms, and have given me the opportunity to learn even more about ROS and (humanoid) robots. Finally, thanks to my family and friends for their help and support.

About the Reviewers

Luis Sánchez Crespo has completed his dual Master's degree in Electronics and Telecommunication Engineering at the University of Las Palmas de Gran Canaria. He has collaborated with different research groups as the Institute for Technological Development and Innovation (IDETIC), the Oceanic Platform of Canary Islands (PLOCAN), and the Institute of Applied Microelectronics (IUMA) where he actually researches on imaging super-resolution algorithms.

His professional interests lie in computer vision, signal processing, and electronic design applied on robotics systems. For this reason, he joined the AVORA team, a group of young engineers and students working on the development of Underwater Autonomous Vehicles (AUV) from scratch. Inside this project, Luis has started developing acoustic and computer vision systems, extracting information from different sensors such as hydrophones, SONAR, or camera. He has also been involved in the electronic design of the vehicle. Finally, he has played the Team Leader role during the preparation of the SAUC-E'13 challenge.

With a strong background gained in marine technology, Luis joined Biomecan, a young startup, where he works on developing remotely operated and autonomous vehicles for aquatic environments.

He is very enthusiastic and an engineer in multiple disciplines. He is responsible for his work. He can manage himself and can take up responsibilities as a Team Leader, as demonstrated at the SAUC-E competition directing the AVORA team. His background in electronics and telecommunications allows him to cover a wide range of expertise from signal processing and software, to electronic design and fabrication.

He has focused his career in 2D and 3D signal processing, with the development of a system for tracking and detecting signs of exhaustion and the risk of falling asleep in drivers. After this successful research, he started working on two different projects at the same time. The first of these projects focused mainly on achieving video sequences enhancement applying super-resolution. The second project, and one of his most important achievements, was participating in the development of an autonomous underwater vehicle for the Students Autonomous Underwater Challenge-Europe (SAUC-E) in which his team achieved great recognition with the fourth most important prize. In his second year, he took up the mantle of Team Leader, again being recognized by his work during competition.

I would like to thank my family for supporting me since my first step, Guaxara for lighting my path, and my teammates for supporting me. I would also like to thank Dario Sosa Cabrera and Anil Motilal Mahtani Mirchandani.

Matthieu Keller is a French student who has completed several internships in development, system administration, and cyber security. His education is mainly in Computer Science and Robotics, but he enjoys all kinds of scientific topics.

Damian Melniczuk graduated with Physics from the Wrocław University of Technology, where he currently works in the quantum cryptography laboratory. Apart from using photons for transporting encryption keys, he is also involved in hacker culture and open source movement. His current projects are: setting up Wroclaw Hackerspace (`http://hswro.org/`) and building an open source modular home automation system (`http://openhomeautomation.blogspot.com/`).

Table of Contents

Preface

Learning ROS for Robotics Programming gives you a comprehensive review of ROS tools. ROS is the Robot Operating System framework, which is used nowadays by hundreds of research groups and companies in the robotics industry. But it is also the painless entry point to robotics for nonprofessional people. You will see how to install ROS, start playing with its basic tools, and you will end up working with state-of-the-art computer vision and navigation tools.

The content of the book can be followed without any special devices, and each chapter comes with a series of source code examples and tutorials that you can run on your own computer. This is the only thing you need to follow in the book. However, we also show you how to work with hardware, so that you can connect your algorithms with the real world. Special care has been taken in choosing devices which are affordable for amateur users, but at the same time the most typical sensors or actuators in robotics research are covered.

Finally, the potential of ROS is illustrated with the ability to work with whole robots in a simulated environment. You will learn how to create your own robot and integrate it with the powerful navigation stack. Moreover, you will be able to run everything in simulation, using the Gazebo simulator. We will end the book by providing a list of real robots available for simulation in ROS. At the end of the book, you will see that you can work directly with them and understand what is going on under the hood.

What this book covers

Chapter 1, Getting Started with ROS, shows the easiest way you must follow in order to have a working installation of ROS. You will see how to install different distributions of ROS, and you will use ROS Fuerte in the rest of the book. How to make an installation from Debian packages or compiling the sources, as well as making installations in virtual machines, have been described in this chapter.

Chapter 2, The ROS Architecture with Examples, is concerned with the concepts and tools provided by the ROS framework. We will introduce you to nodes, topics, and services, and you will also learn how to use them. Through a series of examples, we will illustrate how to debug a node or visualize the messages published through a topic.

Chapter 3, Debugging and Visualization, goes a step further in order to show you powerful tools for debugging your nodes and visualize the information that goes through the node's graph along with the topics. ROS provides a logging API which allows you to diagnose node problems easily. In fact, we will see some powerful graphical tools such as rxconsole and rxgraph, as well as visualization interfaces such as rxplot and rviz. Finally, this chapter explains how to record and playback messages using rosbag and rxbag.

Chapter 4, Using Sensors and Actuators with ROS, literally connects ROS with the real world. This chapter goes through a number of common sensors and actuators that are supported in ROS, such as range lasers, servo motors, cameras, RGB-D sensors, and much more. Moreover, we explain how to use embedded systems with microcontrollers, similar to the widely known Arduino boards.

Chapter 5, 3D Modeling and Simulation, constitutes one of the first steps in order to implement our own robot in ROS. It shows you how to model a robot from scratch and run it in simulation using the Gazebo simulator. This will later allow you to use the whole navigation stack provided by ROS and other tools.

Chapter 6, Computer Vision, shows the support for cameras and computer vision tasks in ROS. This chapter starts with drivers available for FireWire and USB cameras, so that you can connect them to your computer and capture images. You will then be able to calibrate your camera using ROS calibration tools. Later, you will be able to use the image pipeline, which is explained in detail. Then, you will see how to use several APIs for vision and integrate OpenCV. Finally, the installation and usage of a visual odometry software is described.

Chapter 7, Navigation Stack – Robot Setups, is the first of two chapters concerned with the ROS navigation stack. This chapter describes how to configure your robot so that it can be used with the navigation stack. In the same way, the stack is explained, along with several examples.

Chapter 8, Navigation Stack – Beyond Setups, continues the discussion of the previous chapter by showing how we can effectively make our robot navigate autonomously. It will use the navigation stack intensively for that. This chapter shows the great potential of ROS using the Gazebo simulator and rviz to create a virtual environment in which we can build a map, localize our robot, and do path planning with obstacle avoidance.

Chapter 9, Combining Everything – Learn by Doing, builds from the previous chapters and shows a number of robots which are supported in ROS using the Gazebo simulator. In this chapter you will see how to run these robots in simulation and perform several of the tasks learned in the rest of the book, especially those related to the navigation stack.

What you need for this book

This book was written with the intention that almost everybody can follow it and run the source code examples provided with it. Basically, you need a computer with a Linux distribution. Although any Linux distribution should be fine, it is recommended that you use a recent version of Ubuntu. Then you will use ROS Fuerte, which is installed according to the instructions given in *Chapter 1, Getting Started with ROS*. For this distribution of ROS, you will need a version of Ubuntu prior to 12.10 because since this version Fuerte is no longer supported.

Regarding the hardware requirements of your computer, in general any computer or laptop is enough. However, it is advisable to use a dedicated graphic card in order to run the Gazebo simulator. Also, it will be good to have a good number of peripherals, so that you can connect several sensors and actuators, including cameras and Arduino boards.

You will also need Git (the `git-core` Debian package) in order to clone the repository with the source code provided with this book. Similarly, you are expected to have a basic knowledge of the Bash command line, GNU/Linux tools, and some C/C++ programming skills.

Who this book is for

This book is targeted at all robotics developers, from amateurs to professionals. It covers all the aspects involved in a whole robotic system and shows how ROS helps with the task of making a robot really autonomous. Anyone who is learning robotics and has heard about ROS but has never tried it will benefit from this book. Also, ROS beginners will learn advance concepts and tools of this framework. Indeed, even regular users may learn something new from some particular chapters. Certainly, only the first three chapters are intended for new users; so those who already use ROS may skip these ones and go directly to the rest.

Conventions

In this book, you will find a number of styles of text that distinguish between different kinds of information. Here are some examples of these styles, and an explanation of their meanings.

Code words in text are shown as follows: "The `*-ros-pkg` contributed packages are licensed under a variety of open source licenses."

A block of code is set as follows:

```
<package>
  <description brief="short description">
    long description,
  </description>
  <author>Aaron Martinez, Enrique Fernandez</author>
  <license>BSD</license>
  <url>http://example.com/</url>

  <depend package="roscpp"/>
  <depend package="common"/>
  <depend package="otherPackage"/>
  <versioncontrol type="svn" url="https://urlofpackage/trunk"/>
  <export>
    <cpp cflags="-I${prefix}/include" lflags="-L${prefix}/lib -lros"/>
</package>
```

When we wish to draw your attention to a particular part of a code block, the relevant lines or items are set in bold:

```
<launch>
<node pkg="chapter3_tutorials" type="example1"
            name="example1" output="screen"
            launch-prefix="xterm -e gdb --args"/>
</launch>
```

Any command-line input or output is written as follows:

```
$ rosrun book_tutorials tutorialX _param:=9.0
```

New terms and **important words** are shown in bold. Words that you see on the screen, in menus, or dialog boxes for example, appear in the text like this: "We must have clicked on the **Play** button at least once."

Warnings or important notes appear in a box like this.

Tips and tricks appear like this.

Reader feedback

Feedback from our readers is always welcome. Let us know what you think about this book—what you liked or may have disliked. Reader feedback is important for us to develop titles that you really get the most out of.

To send us general feedback, simply send an e-mail to feedback@packtpub.com, and mention the book title via the subject of your message.

If there is a topic that you have expertise in and you are interested in either writing or contributing to a book, see our author guide on www.packtpub.com/authors.

Customer support

Now that you are the proud owner of a Packt book, we have a number of things to help you to get the most from your purchase.

Downloading the example code

You can download the example code files for all Packt books you have purchased from your account at http://www.packtpub.com. If you purchased this book elsewhere, you can visit http://www.packtpub.com/support and register to have the files e-mailed directly to you. You can also download these code files from https://github.com/AaronMR/Learning_ROS_for_Robotics_Programming.

Downloading the color images of this book

We also provide you a PDF file that has color images of the screenshots/diagrams used in this book. The color images will help you better understand the changes in the output. You can download this file from http://www.packtpub.com/sites/default/files/downloads/1448OS_Graphics.pdf.

Errata

Although we have taken every care to ensure the accuracy of our content, mistakes do happen. If you find a mistake in one of our books—maybe a mistake in the text or the code—we would be grateful if you would report this to us. By doing so, you can save other readers from frustration and help us improve subsequent versions of this book. If you find any errata, please report them by visiting `http://www.packtpub.com/submit-errata`, selecting your book, clicking on the **errata submission form** link, and entering the details of your errata. Once your errata are verified, your submission will be accepted and the errata will be uploaded on our website, or added to any list of existing errata, under the Errata section of that title. Any existing errata can be viewed by selecting your title from `http://www.packtpub.com/support`.

Piracy

Piracy of copyright material on the Internet is an ongoing problem across all media. At Packt, we take the protection of our copyright and licenses very seriously. If you come across any illegal copies of our works, in any form, on the Internet, please provide us with the location address or website name immediately so that we can pursue a remedy.

Please contact us at `copyright@packtpub.com` with a link to the suspected pirated material.

We appreciate your help in protecting our authors, and our ability to bring you valuable content.

Questions

You can contact us at `questions@packtpub.com` if you are having a problem with any aspect of the book, and we will do our best to address it.

1
Getting Started with ROS

Welcome to the first chapter of this book where you will learn how to install ROS, the new standard software framework in robotics. With ROS, you will start to program and control your robots the easy way using tons of examples and source code that will show you how to use sensors and devices or add new functionalities to your robot, such as autonomous navigation and visual perception. Thanks to the open source motto and the community that is developing the state-of-the-art algorithms and providing new functionalities, ROS is growing every day.

In this book you will learn the following:

- Installing the ROS framework on a version of Ubuntu
- Learning the basic operation of ROS
- Debugging and visualizing the data
- Programming your robot using this framework
- Creating a 3D model to use it in the simulator
- Using the navigation stack to make your robot autonomous

In this chapter we are going to install a full version of ROS in Ubuntu. We will use Ubuntu because it is fully supported by and recommended for ROS. However, you can use a different operating system instead of Ubuntu, but in these operative systems, ROS is still experimental and could have some errors. So, for this reason, we recommend you to use Ubuntu while you follow the samples in this book.

Before starting with the installation, we are going to learn the origin of ROS and its history.

Robot Operating System (**ROS**) is a framework that is widely used in robotics. The philosophy is to make a piece of software that could work in other robots by making little changes in the code. What we get with this idea is to create functionalities that can be shared and used in other robots without much effort so that we do not reinvent the wheel.

ROS was originally developed in 2007 by the Stanford Artificial Intelligence Laboratory (SAIL) with the support of the Stanford AI Robot project. As of 2008, development continues primarily at Willow Garage, a robotics research institute, with more than 20 institutions collaborating within a federated development model.

A lot of research institutions have started to develop projects in ROS by adding hardware and sharing their code samples. Also, the companies have started to adapt their products to be used in ROS. In the following image, you can see some fully supported platforms. Normally, these platforms are published with a lot of code, examples, and simulators to permit the developers to start their work easily.

The sensors and actuators used in robotics have also been adapted to be used with ROS. Every day an increasing number of devices are supported by this framework.

ROS provides standard operating system facilities such as hardware abstraction, low-level device control, implementation of commonly used functionalities, message passing between processes, and package management. It is based on graph architecture with a centralized topology where processing takes place in nodes that may receive or post, such as multiplex sensor, control, state, planning, actuator, and so on. The library is geared towards a Unix-like system (Ubuntu Linux is listed as supported while other variants such as Fedora and Mac OS X are considered experimental).

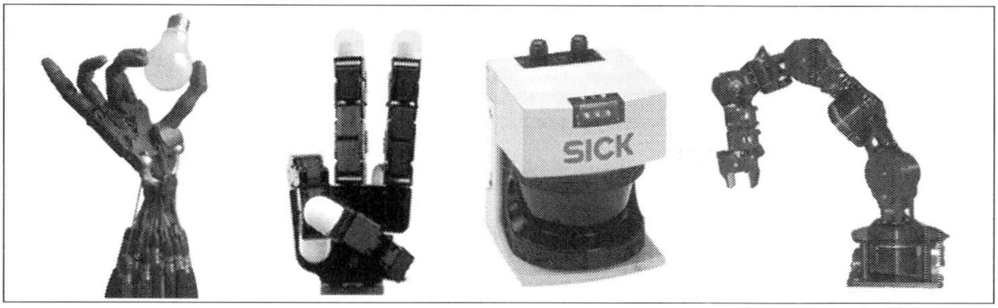

The `*-ros-pkg` package is a community repository for developing high-level libraries easily. Many of the capabilities frequently associated with ROS, such as the navigation library and the rviz visualizer, are developed in this repository. These libraries give a powerful set of tools to work with ROS easily, knowing what is happening every time. Of these, visualization, simulators, and debugging tools are the most important ones.

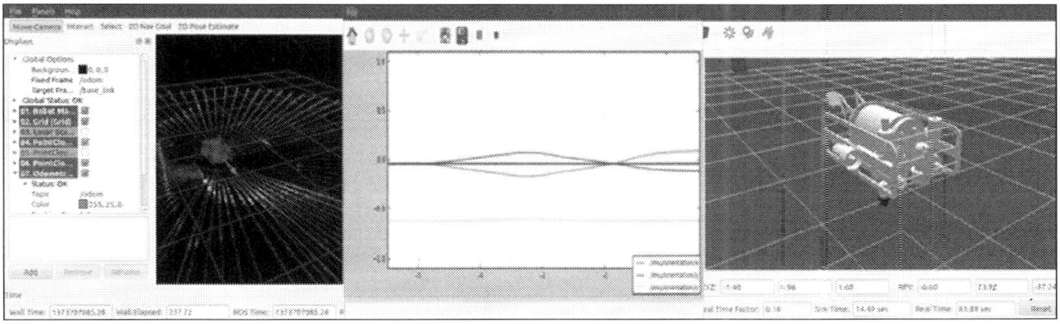

ROS is released under the terms of the **BSD** (**Berkeley Software Distribution**) license and is an open source software. It is free for commercial and research use. The `*-ros-pkg` contributed packages are licensed under a variety of open source licenses.

ROS promotes code reutilization so that the robotics developers and scientists do not have to reinvent the wheel all the time. With ROS, you can do this and more. You can take the code from the repositories, improve it, and share it again.

ROS has released some versions, the latest one being Groovy. In this book, we are going to use Fuerte because it is a stable version, and some tutorials and examples used in this book don't work in the Groovy version.

Now we are going to show you how to install ROS Electric and Fuerte. Although in this book we use Fuerte, you may need to install the Electric version to use some code that works only in this version or you may need Electric because your robot doesn't have the latest version of Ubuntu.

As we said before, the operating system used in the book is Ubuntu and we are going to use it in all tutorials. If you are using another operating system and you want to follow the book, the best option is to install a virtual machine with an Ubuntu copy. Later, we will explain how to install a virtual machine in order to use ROS in it.

Anyway, if you want to try installing it in an operating system other than Ubuntu, you can find the required instructions in the following link: `http://wiki.ros.org/fuerte/Installation`.

Installing ROS Electric – using repositories

There are a few methods available to install ROS. You can do it directly using repositories, the way we will do now, or you can use the code files and compile it. It is more secure to do it using repositories because you have the certainty that it will work.

In this section, you will see the steps to install ROS Electric on your computer. The installation process has been explained in detail in the official ROS page: `http://wiki.ros.org/electric/Installation`.

We assume that you know what an Ubuntu repository is and how to manage it. If you have any queries, check the following link to get more information: `https://help.ubuntu.com/community/Repositories/Ubuntu`.

Before starting with the installation, we need to configure our repositories. To do this, the repositories need to allow restricted, universal, and multiversal repositories. To check whether your version of Ubuntu accepts these repositories, click on **Ubuntu Software Center** in the menu on the left of your desktop.

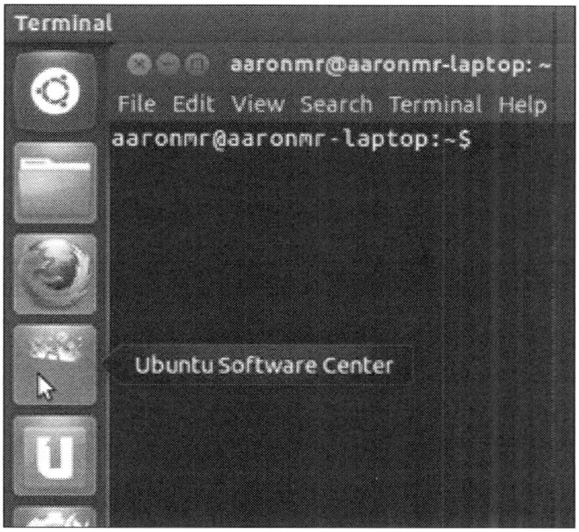

Navigate to **Edit | Software Sources** and you will see the following window on your screen. Make sure that everything is selected as shown in the following screenshot:

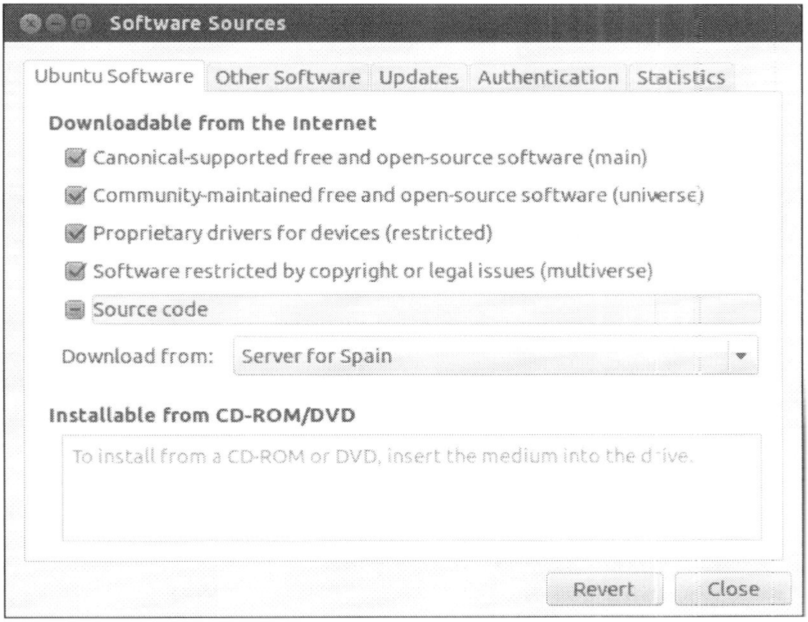

Normally, these options are marked, so you will not have problems with this step.

Adding repositories to your sources.list file

In this step, you have to select your Ubuntu version. It is possible to install ROS Electric in various versions of the operating system. You can use any of them, but we recommend you to always use the most updated version to avoid problems:

- A specific way to install the repositories for an Ubuntu-based distro such as Ubuntu Lucid Lynx (10.04) is as follows:

  ```
  $ sudo sh -c 'echo "deb http://packages.ros.org/ros/ubuntu lucid main" > /etc/apt/sources.list.d/ros-latest.list'
  ```

- A generic way for installing any distro of Ubuntu relies on the `lsb_release` command that is supported on all Linux Debian-based distro:

  ```
  $ sudo sh -c 'echo "deb http://packages.ros.org/ros/ubuntu `lsb_release -cs` main" > /etc/apt/sources.list.d/ros-latest.list'
  ```

Once you have added the correct repository, your operating system knows where to download the programs that need to be installed on your system.

Setting up your keys

This step is to confirm that the origin of the code is correct, and nobody has modified the code or programs without the knowledge of the owner. Normally, when you add a new repository, you have to add the keys of that repository so that it is added to your system's trusted list:

```
$ wget http://packages.ros.org/ros.key -O - | sudo apt-key add -
```

We can now be sure that the code came from an authorized site.

Installation

Now we are ready to start the installation. Before we start, it would be better to update the software to avoid problems with libraries or the wrong software version. We do this with the following command:

```
$ sudo apt-get update
```

ROS is huge; sometimes you will install libraries and programs that you will never use. Normally, it has four different installations depending on the final use; for example, if you are an advanced user, you may only need basic installation for a robot without enough space in the hard disk. For this book, we recommend the use of full installation because it will install everything that's necessary to make the examples and tutorials work.

Don't worry if you don't know what you are installing right now, be it rviz, simulators, or navigation. You will learn everything in the upcoming chapters:

- The easiest (and recommended if you have enough hard disk space) installation is known as desktop-full. It comes with ROS, the Rx tools, the rviz visualizer (for 3D), many generic robot libraries, the simulator in 2D (such as stage) and 3D (usually Gazebo), the navigation stack (to move, localize, do mapping, and control arms), and also perception libraries using vision, lasers or RGB-D cameras:

  ```
  $ sudo apt-get install ros-electric-desktop-full
  ```

- If you do not have enough disk space, or you prefer to install only a few stacks, first install only the desktop installation file, which comes only with ROS, the Rx tools, rviz, and generic robot libraries. Later, you can install the rest of the stacks when you need them (using aptitude and looking for the `ros-electric-*` stacks, for example):

  ```
  $ sudo apt-get install ros-electric-desktop
  ```

- If you only want the bare bones, install ROS-base, which is usually recommended for the robot itself or computers without a screen or just a TTY. It will install the ROS package with the build and communication libraries and no GUI tools at all:

  ```
  $ sudo apt-get install ros-electric-ros-base
  ```

- Finally, along with whatever option you choose from this list, you can install individual/specific ROS stacks (for a given stack name):

  ```
  $ sudo apt-get install ros-electric-STACK
  ```

The environment setup

Congratulations! You are in this step because you have an installed version of ROS on your system. To start using it, the system must know where the executable or binary files as well as other commands are. To do this, you need to execute the next script. If you install another ROS distro in addition to your existing version, you can work with both by calling the script of the one you need every time, since this script simply sets your environment. Here, we will use the one for ROS Electric, but just change `electric` to `fuerte` or `groovy` in the following command if you want to try other distros:

```
$ source /opt/ros/electric/setup.bash
```

If you type `roscore` in the shell, you will see that something is starting. This is the best test to find out whether you have ROS and whether it is installed correctly.

Note that if you open another shell and type `roscore` or any other ROS command, it does not work. This is because it is necessary to execute the script again to configure the global variables and path for the location where ROS is installed.

It is very easy to solve this. You only need to add the script at the end of your `.bashrc` file and when you start a new shell, the script will execute and you will have the environment configured. Use the following command to do this:

```
$ echo "source /opt/ros/electric/setup.bash" >> ~/.bashrc
$ source ~/.bashrc
```

If it happens that you have more than a single ROS distribution installed on your system, your `~/.bashrc` file must source only `setup.bash` of the version you are currently using. This is because the last call will override the environment set of the others, as we have mentioned previously, to have several distros living in the same system and switch among them.

Installing ROS Fuerte – using repositories

In this section, we are going to install ROS Fuerte on our computer. You can have different versions installed on the same computer without problems; you only need to select the version that you want to use in the `.bashrc` file. You will see how to do this in this section.

If you want to see the official page where this process is explained, you can visit the following URL: `http://wiki.ros.org/fuerte/Installation`.

You can install ROS using two methods: using repositories and using source code. Normal users will only need to make an installation using repositories to get a functional installation of ROS. You can install ROS using the source code but this process is for advanced users and we don't recommend it.

Configuring your Ubuntu repositories

First, you must check that your Ubuntu accepts restricted, universal, and multiversal repositories. Refer to the *Installing ROS Electric – using repositories* section if you want to see how to do it.

Normally, Ubuntu is configured to allow these repositories and you won't have problems with this step.

Setting up your source.list file

Now we are going to add the URLs from where we can download the code. Note that ROS Fuerte doesn't work for Maverick and Natty, so you must have Ubuntu 10.04, 11.10, or 12.04 on your computer.

For this book we have used Ubuntu 12.04 and it works fine. All the examples have been checked, compiled, and executed in this version of Ubuntu.

Open a new shell and type the following command, as we did before, which should work for any Ubuntu version you have:

```
$ sudo sh -c 'echo "deb http://packages.ros.org/ros/ubuntu `lsb_release -cs` main" > /etc/apt/sources.list.d/ros-latest.list'
```

Setting up your keys

It is important to add the key because with it we can be sure that we are downloading the code from the right place and nobody has modified it.

If you have followed the steps to install ROS Electric, you don't need to do this again as you have already completed this earlier; if not, add the repository using the following command:

```
$ wget http://packages.ros.org/ros.key -O - | sudo apt-key add -
```

Installation

We are ready to install ROS Fuerte at this point. Before doing something, it is necessary to update all the programs used by ROS. We do it to avoid incompatibility problems.

Type the following command in a shell and wait:

```
$ sudo apt-get update
```

Depending on whether you had the system updated or not, the command will take more or less time to finish.

ROS has a lot of parts and installing the full system can be heavy in robots without enough features. For this reason, you can install various versions depending on what you want to install.

For this book, we are going to install the full version. This version will install all the examples, stacks, and programs. This is a good option for us because in some chapters of this book, we will need to use tools, and if we don't install it now, we will have to do it later:

- The easiest (and recommended if you have enough hard disk space) installation is known as desktop-full. It comes with ROS, the Rx tools, the rviz visualizer (for 3D), many generic robot libraries, the simulator in 2D (such as stage) and 3D (usually Gazebo), the navigation stack (to move, localize, do mapping, and control arms), and also perception libraries using vision, lasers, or RGB-D cameras:

  ```
  $ sudo apt-get install ros-fuerte-desktop-full
  ```

- If you do not have enough disk space, or you prefer to install only a few stacks, first install only the desktop installation file, which comes only with ROS, the Rx tools, rviz, and generic robot libraries. Later, you can install the rest of the stacks when you need them (using aptitude and looking for the `ros-electric-*` stacks, for example):

  ```
  $ sudo apt-get install ros-fuerte-desktop
  ```

- If you only want the bare bones, install ROS-comm, which is usually recommended for the robot itself or computers without a screen or just a TTY. It will install the ROS package with the build and communication libraries and no GUI tools at all:

  ```
  $ sudo apt-get install ros-fuerte-ros-comm
  ```

- Finally, along with whatever option you choose from the list, you can install individual/specific ROS stacks (for a given stack name):

  ```
  $ sudo apt-get install ros-fuerte-STACK
  ```

Do not worry if you are installing things that you do not know. In the upcoming chapters, you will learn about everything you are installing and how to use it.

When you gain experience with ROS, you can make basic installations in your robots using only the core of ROS, using less resources, and taking only what you need.

> **Downloading the example code**
>
> You can download the example code files for all Packt books you have purchased from your account at `http://www.packtpub.com`. If you purchased this book elsewhere, you can visit `http://www.packtpub.com/support` and register to have the files e-mailed directly to you. You can also download these code files from `https://github.com/AaronMR/Learning_ROS_for_Robotics_Programming`.

The environment setup

Now that you have installed ROS, to start using it, you must provide Ubuntu with the path where ROS is installed. Open a new shell and type the following command:

```
$ roscore
roscore: command not found
```

You will see this message because Ubuntu does not know where to search for the commands. To solve it, type the following command in a shell:

```
$ source /opt/ros/fuerte/setup.bash
```

Then, type the `roscore` command once again, and you will see the following output:

```
...
started roslaunch server http://localhost:45631/
ros_comm version 1.8.11

SUMMARY
========

PARAMETERS
* /rosdistro
* /rosversion

NODES

auto-starting new master
....
```

This means that Ubuntu knows where to find the commands to run ROS. Note that if you open another shell and type `roscore`, it will not work. This is because it is necessary to add this script within the `.bashrc` file. So, every time you start a new shell, the scripts will run because `.bashrc` always runs when a shell runs.

Use the following commands to add the script:

```
$ echo "source /opt/ros/fuerte/setup.bash" >> ~/.bashrc
$ source ~/.bashrc
```

As mentioned before, only source `setup.bash` for one ROS distribution. Imagine that you had Electric and Fuerte installed on your computer, and you are using Fuerte as the normal version. If you want to change the version used in a shell, you only have to type the following command:

```
$ source /opt/ros/electric/setup.bash
```

If you want to use another version on a permanent basis, you must change the `.bashrc` file and put the correct script for your version.

Standalone tools

ROS has some tools that need to be installed after the principal installation. These tools will help us install dependencies between programs to compile, download, and install packages from ROS. These tools are `rosinstall` and `rosdep`. We recommend installing them because we will use them in the upcoming chapters. To install these tools, type the following command in a shell:

```
$ sudo apt-get install python-rosinstall python-rosdep
```

Now we have a full installation of ROS on our system. As you can see, only a few steps are necessary to do it.

It is possible to have two or more versions of ROS installed on our computer. Furthermore, you can install ROS on a virtual machine if you don't have Ubuntu installed on your computer.

In the next section, we will explain how to install a virtual machine and use a drive image with ROS. Perhaps this is the best way to get ROS for new users.

How to install VirtualBox and Ubuntu

VirtualBox is a general-purpose, full virtualizer for x86 hardware, targeted at server, desktop, and embedded use. VirtualBox is free and supports all the major operating systems and pretty much every Linux flavor out there.

As we recommend the use of Ubuntu, you perhaps don't want to change the operating system of your computer. Tools such as VirtualBox exist for this purpose and help us virtualize a new operating system on our computer without making any changes to the original.

In the upcoming sections, we are going to show how to install VirtualBox and a new installation of Ubuntu. Further, with this virtual installation, you could have a clean installation to restart your development machine if you have any problem, or to save the development machine with all the setup files necessary for your robot.

Downloading VirtualBox

The first step is to download the VirtualBox installation file. At the time of writing, the following links had the latest versions available:

- `https://www.virtualbox.org/wiki/Downloads`
- `http://download.virtualbox.org/virtualbox/4.2.0/VirtualBox-4.2.1-80871-OSX.dmg`

Once installed, you need to download the image of Ubuntu. For this tutorial, we will use an Ubuntu copy with ROS Fuerte installed. You can download it from the following URL: `http://nootrix.com/wp-content/uploads/2012/08/ROS.ova`.

You can find different virtual machines with preinstalled Ubuntu and ROS, but we are going to use this version because it is referred by the official page of ROS.

Creating the virtual machine

Creating a new virtual machine with the downloaded file is very easy; just follow the steps outlined in this section. Open VirtualBox and navigate to **File | Import Appliance...**. Then, click on **Open appliance** and select the `ROS.ova` file downloaded before.

In the next window, you can configure the parameters for the new virtual machine. Keep the default configuration and only change the name of the virtual system. This name helps you distinguish this virtual machine from others. Our recommendation is to put a descriptive name; in our case, the book's name.

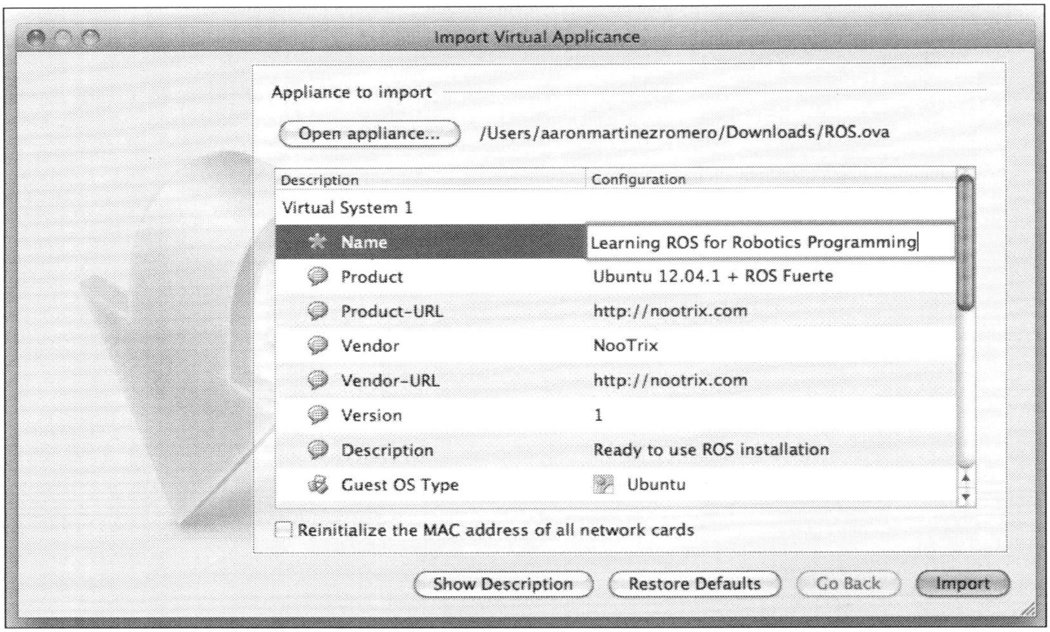

Click on the **Import** button and accept the software license agreement in the next window. Then, you will see a progress bar. It means that VirtualBox is copying the file with the virtual image and is creating a new copy with the new name.

Note that this process doesn't modify the original file `ROS.ova`, and you could create more virtual machines with different copies from the original file.

The process will take a few minutes depending on the speed of your computer. When it finishes, you can start your virtual machine by clicking on the **Start** button. Remember to select the right machine before starting it. In our case, we have only one machine but you could have more.

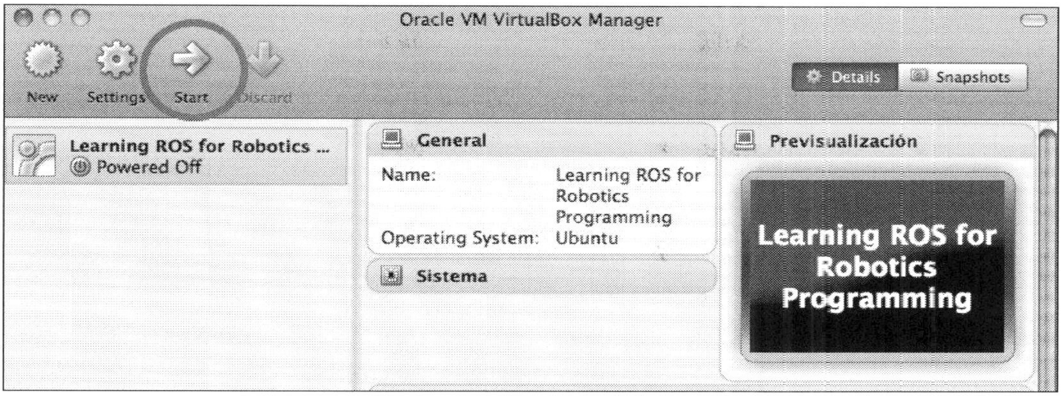

Sometimes you will get the error shown in the following screenshot. It is because your computer doesn't have the correct drivers to use USB 2.0. You can fix it by installing Oracle VM VirtualBox Extension Pack, but you can also disable the USB support to start using the virtual machine.

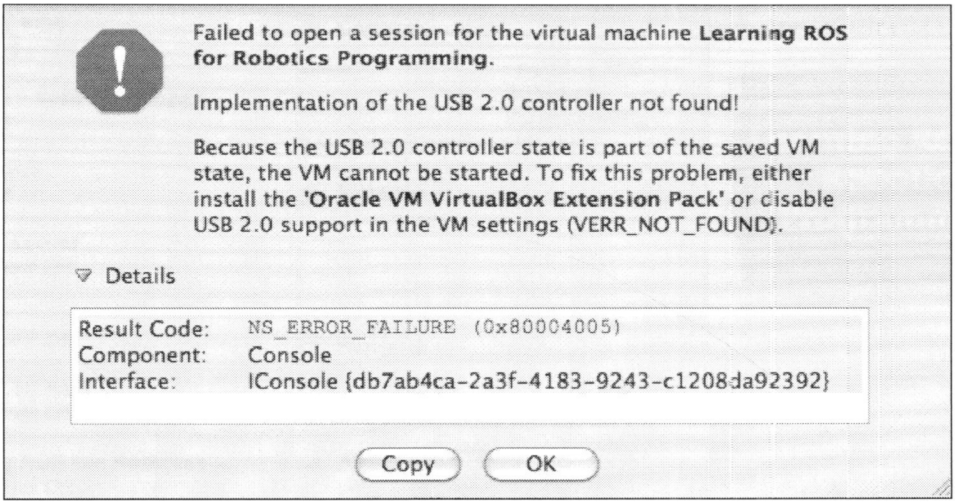

To disable USB support, right-click on the virtual machine and select **Settings**. In the toolbar, navigate to **Ports** | **USB** and uncheck **Enable USB 2.0 (EHCI) Controller**. You can now start the virtual machine again, and it should start without problems.

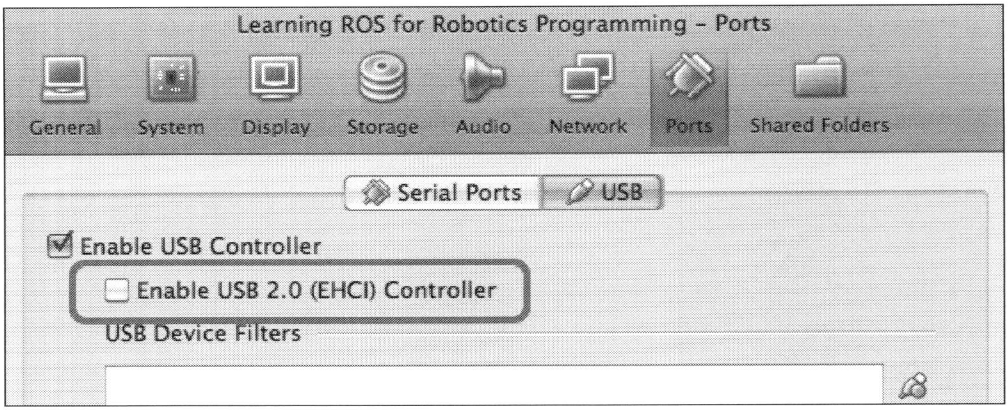

Once the virtual machine starts, you should see the ROS-installed Ubuntu 12.04 window on your screen as shown in the following screenshot:

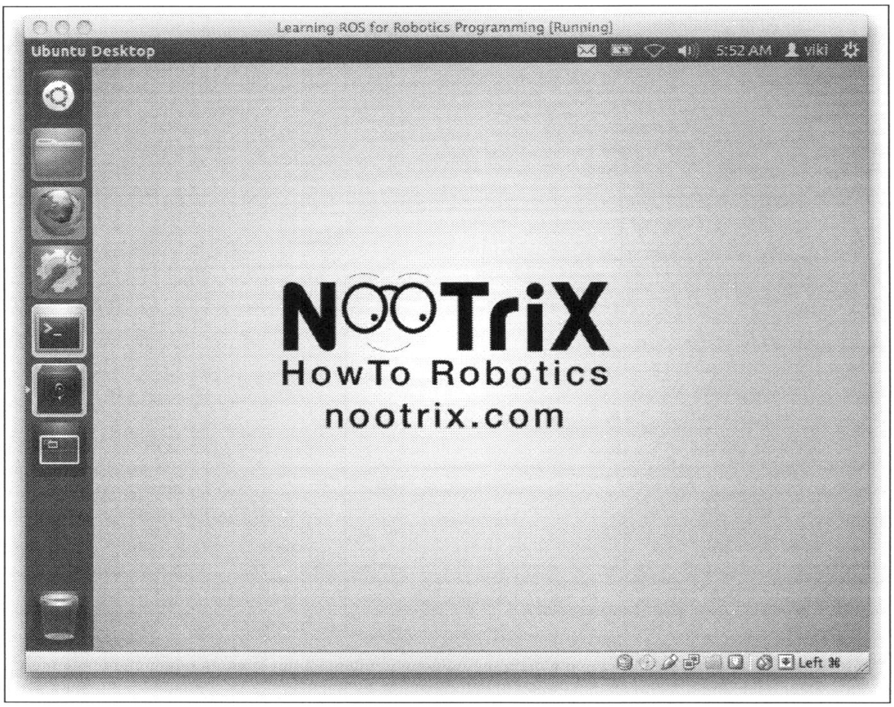

When you have finished these steps, you will have a full copy of ROS Fuerte that can be used in this book. You can run all the examples and stacks that we are going to work with. Unfortunately, VirtualBox has problems while working with real hardware, and it's possible that you can't use this copy of ROS Fuerte for the steps outlined in *Chapter 4, Using Sensors and Actuators with ROS.*

Summary

In this chapter we have learned to install two different versions of ROS (Electric and Fuerte) in Ubuntu. With these steps, you have all the necessary software installed on your system to start working with ROS and the examples of this book. You can install ROS using the source code as well. This option is for advanced users, and we recommend that you use this repository only for installation; it is more common and normally it should not give errors or problems.

It is a good idea to play with ROS and the installation on a virtual machine. This way, if you have problems with the installation or with something, you can reinstall a new copy of your operating system and start again.

Normally, with virtual machines, you will not have access to real hardware, such as sensors and actuators. Anyway, you can use it for testing algorithms, for example.

In the next chapter, you will learn the architecture of ROS, some important concepts, and some tools to interact directly with ROS.

2

The ROS Architecture with Examples

Once you have installed ROS, you surely must be thinking, "OK, I have installed it, and now what?" In this chapter, you will learn the structure of ROS and what parts it has. Furthermore, you will start to create nodes, packages, and use ROS with examples using TurtleSim.

The ROS architecture has been designed and divided into three sections or levels of concepts:

- The Filesystem level
- The Computation Graph level
- The Community level

The first level is the Filesystem level. In this level, a group of concepts is used to explain how ROS is internally formed, the folder structure, and the minimal files that it needs to work.

The second level is the Computation Graph level where communication between processes and systems happen. In this section, we will see all the concepts and systems that ROS has to set up systems, to handle all the processes, to communicate with more than a single computer, and so on.

The third level is the Community level where we will explain the tools and concepts to share knowledge, algorithms, and code from any developer. This level is important because ROS can grow quickly with great support from the community.

Understanding the ROS Filesystem level

When you start to use or develop projects with ROS, you will start to see this concept that could sound strange in the beginning, but as you use ROS, it will begin to become familiar to you.

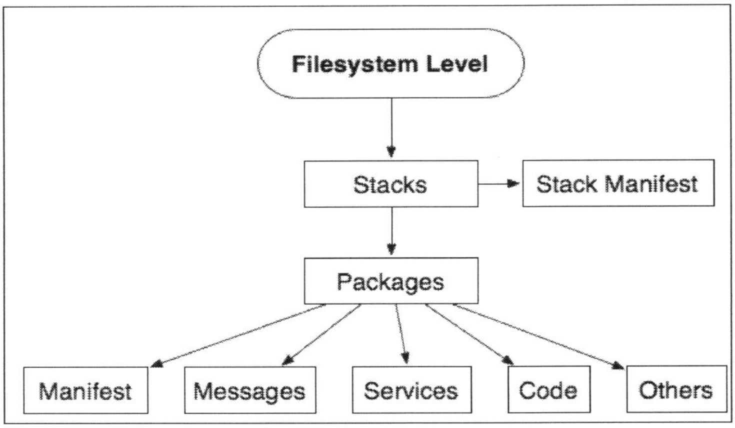

Similar to an operating system, an ROS program is divided into folders, and these folders have some files that describe their functionalities:

- **Packages**: Packages form the atomic level of ROS. A package has the minimum structure and content to create a program within ROS. It may have ROS runtime processes (nodes), configuration files, and so on.
- **Manifests**: Manifests provide information about a package, license information, dependencies, compiler flags, and so on. Manifests are managed with a file called `manifests.xml`.
- **Stacks**: When you gather several packages with some functionality, you will obtain a stack. In ROS, there exists a lot of these stacks with different uses, for example, the navigation stack.
- **Stack manifests**: Stack manifests (`stack.xml`) provide data about a stack, including its license information and its dependencies on other stacks.
- **Message (msg) types**: A message is the information that a process sends to other processes. ROS has a lot of standard types of messages. Message descriptions are stored in `my_package/msg/MyMessageType.msg`.
- **Service (srv) types**: Service descriptions, stored in `my_package/srv/MyServiceType.srv`, define the request and response data structures for services in ROS.

In the following screenshot, you can see the contents of the `chapter3` folder of this book. What you see is a package where the examples of code for the chapter are stored along with the manifest, and much more. The `chapter3` folder does not have messages and services, but if it had, you would have seen the `srv` and `msg` folders.

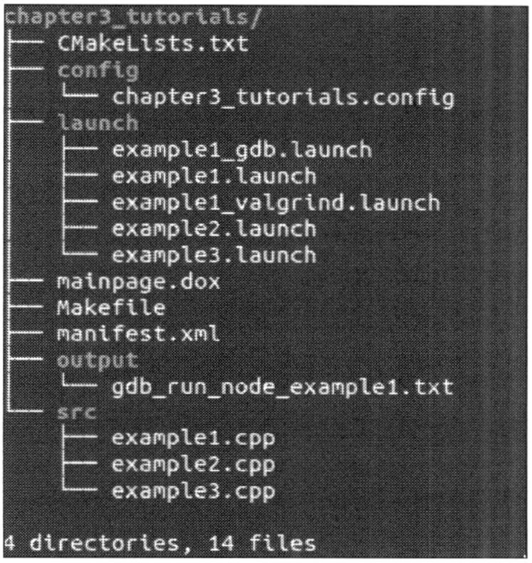

Packages

Usually, when we talk about packages, we refer to a typical structure of files and folders. This structure looks as follows:

- `bin/`: This is the folder where our compiled and linked programs are stored after building/making them.
- `include/package_name/`: This directory includes the headers of libraries that you would need. Do not forget to export the manifest, since they are meant to be provided for other packages.
- `msg/`: If you develop nonstandard messages, put them here.
- `scripts/`: These are executable scripts that can be Bash, Python, or any other script.
- `src/`: This is where the source files of your programs are present. You may create a folder for nodes and nodelets, or organize it as you want.
- `srv/`: This represents the service (`srv`) types.
- `CMakeLists.txt`: This is the CMake build file.
- `manifest.xml`: This is the package manifest file.

To create, modify, or work with packages, ROS gives us some tools for assistance:

- `rospack`: This command is used to get information or find packages in the system
- `roscreate-pkg`: When you want to create a new package, use this command
- `rosmake`: This command is used to compile a package
- `rosdep`: This command installs system dependencies of a package
- `rxdeps`: This command is used if you want to see the package dependencies as a graph

To move between the packages and their folders and files, ROS gives us a very useful package called `rosbash`, which provides some commands very similar to the Linux commands. The following are a few examples:

- `roscd`: This command helps to change the directory; this is similar to the `cd` command in Linux
- `rosed`: This command is used to edit a file
- `roscp`: This command is used to copy a file from some package
- `rosd`: This command lists the directories of a package
- `rosls`: This command lists the files from a package; this is similar to the `ls` command in Linux

The file `manifest.xml` must be in a package, and it is used to specify information about the package. If you find this file inside a folder, probably this folder is a package.

If you open a `manifest.xml` file, you will see information about the name of the package, dependencies, and so on. All of this is to make the installation and the distribution of these packages easy.

Two typical tags that are used in the manifest file are `<depend>` and `<export>`. The `<depend>` tag shows which packages must be installed before installing the current package. This is because the new package uses some functionality of the other package. The `<export>` tag tells the system what flags need to be used to compile the package, what headers are to be included, and so on.

The following code snippet is an example of this file:

```
<package>
  <description brief="short description">
    long description,
  </description>
  <author>Aaron Martinez, Enrique Fernandez</author>
  <license>BSD</license>
  <url>http://example.com/</url>
```

```
    <depend package="roscpp"/>
    <depend package="common"/>
    <depend package="otherPackage"/>
    <versioncontrol type="svn" url="https://urlofpackage/trunk"/>
    <export>
      <cpp cflags="-I${prefix}/include" lflags="-L${prefix}/lib -lros"/>
</package>
```

Stacks

Packages in ROS are organized into ROS stacks. While the goal of packages is to create minimal collections of code for easy re-use, the goal of stacks is to simplify the process of code sharing.

A stack needs a basic structure of files and folders. You can create it manually, but ROS provides us with the command tool `roscreate-stack` for this process.

The following three files are necessary for a stack: `CMakeList.txt`, `Makefile`, and `stack.xml`. If you see `stack.xml` in a folder, you can be sure that this is a stack.

The following code snippet is an example of this file:

```
<stack>
  <description brief="Sample_Stack">Sample_Stack1</description>
  <author>Maintained by AaronMR</author>
  <license>BSD,LGPL,proprietary</license>
  <review status="unreviewed" notes=""/>
  <url>http://someurl.blablabla</url>
  <depend stack="common_msgs" /> <!-- nav_msgs, sensor_msgs, geometry_
msgs -->
  <depend stack="ros_tutorials" /> <!-- turtlesim -->
</stack>
```

Messages

ROS uses a simplified message description language for describing the data values that ROS nodes publish. With this description, ROS can generate the right source code for these types of messages in several programming languages.

ROS has a lot of messages predefined, but if you develop a new message, it will be in the `msg/` folder of your package. Inside that folder, some files with the `.msg` extension define the messages.

A message must have two principal parts: fields and constants. Fields define the type of data to be transmitted in the message, for example, `int32`, `float32`, and `string`, or new types that you have created before, such as `type1` and `type2`. Constants define the name of the fields.

An example of an `msg` file is as follows:

```
int32 id
float32 vel
string name
```

In ROS, you can find a lot of standard types to use in messages as shown in the following table:

Primitive type	Serialization	C++	Python
`bool`	Unsigned 8-bit int	`uint8_t`	`bool`
`int8`	Signed 8-bit int	`int8_t`	`int`
`uint8`	Unsigned 8-bit int	`uint8_t`	`int`
`int16`	Signed 16-bit int	`int16_t`	`int`
`uint16`	Unsigned 16-bit int	`uint16_t`	`int`
`int32`	Signed 32-bit int	`int32_t`	`int`
`uint32`	Unsigned 32-bit int	`uint32_t`	`int`
`int64`	Signed 64-bit int	`int64_t`	`long`
`uint64`	Unsigned 64-bit int	`uint64_t`	`long`
`float32`	32-bit IEEE float	`float`	`float`
`float64`	64-bit IEEE float	`double`	`float`
`string`	ASCII string (4-bit)	`std::string`	`string`
`time`	Secs/nsecs signed 32-bit ints	`ros::Time`	`rospy.Time`
`duration`	Secs/nsecs signed 32-bit ints	`ros::Duration`	`rospy.Duration`

Downloading the example code

You can download the example code files for all Packt books you have purchased from your account at `http://www.packtpub.com`. If you purchased this book elsewhere, you can visit `http://www.packtpub.com/support` and register to have the files e-mailed directly to you. You can also download these code files from `https://github.com/AaronMR/Learning_ROS_for_Robotics_Programming`.

A special type in ROS is `Header`. This is used to add the timestamp, frame, and so on. This allows messages to be numbered so that we can know who is sending the message. Other functions can be added, which are transparent to the user but are being handled by ROS.

The `Header` type contains the following fields:

```
uint32 seq
time stamp
string frame_id
```

Thanks to `Header`, it is possible to record the timestamp and frame of what is happening with the robot, as we will see in the upcoming chapters.

In ROS, there exist some tools to work with messages. The tool `rosmsg` prints out the message definition information and can find the source files that use a message type.

In the upcoming sections, we will see how to create messages with the right tools.

Services

ROS uses a simplified service description language for describing ROS service types. This builds directly upon the ROS `msg` format to enable request/response communication between nodes. Service descriptions are stored in `.srv` files in the `srv/` subdirectory of a package.

To call a service, you need to use the package name along with the service name; for example, for the file `sample_package1/srv/sample1.srv`, you will refer to it as `sample_package1/sample1`.

There exist some tools that perform some functions with services. The tool `rossrv` prints out service descriptions, packages that contain the `.srv` files, and can find source files that use a service type.

If you want to create a service, ROS can help you with the services generator. These tools generate code from an initial specification of the service. You only need to add the line `gensrv()` to your `CMakeLists.txt` file.

In the upcoming sections, we will learn how to create our own services.

Understanding the ROS Computation Graph level

ROS creates a network where all the processes are connected. Any node in the system can access this network, interact with other nodes, see the information that they are sending, and transmit data to the network.

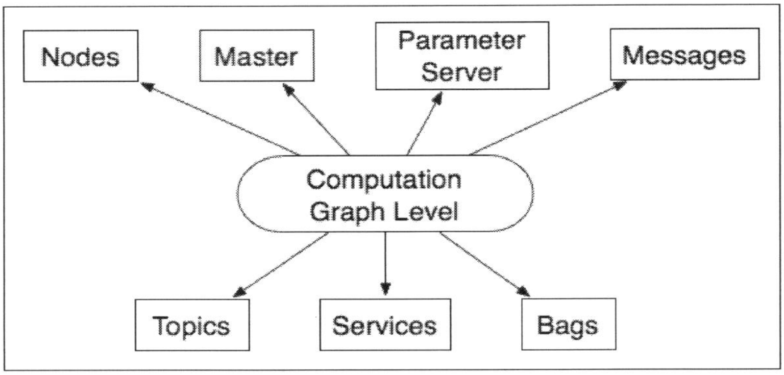

The basic concepts in this level are nodes, the Master, the Parameter Server, messages, services, topics, and bags, all of which provide data to the graph in different ways:

- **Nodes**: Nodes are processes where computation is done. If you want to have a process that can interact with other nodes, you need to create a node with this process to connect it to the ROS network. Usually, a system will have many nodes to control different functions. You will see that it is better to have many nodes that provide only a single functionality, rather than a large node that makes everything in the system. Nodes are written with an ROS client library, for example, `roscpp` or `rospy`.
- **Master**: The Master provides name registration and lookup for the rest of the nodes. If you don't have it in your system, you can't communicate with nodes, services, messages, and others. But it is possible to have it in a computer where nodes work in other computers.
- **Parameter Server**: The Parameter Server gives us the possibility to have data stored using keys in a central location. With this parameter, it is possible to configure nodes while it's running or to change the working of the nodes.
- **Messages**: Nodes communicate with each other through messages. A message contains data that sends information to other nodes. ROS has many types of messages, and you can also develop your own type of message using standard messages.

- **Topics**: Each message must have a name to be routed by the ROS network. When a node is sending data, we say that the node is publishing a topic. Nodes can receive topics from other nodes simply by subscribing to the topic. A node can subscribe to a topic, and it isn't necessary that the node that is publishing this topic should exist. This permits us to decouple the production of the consumption. It's important that the name of the topic must be unique to avoid problems and confusion between topics with the same name.
- **Services**: When you publish topics, you are sending data in a many-to-many fashion, but when you need a request or an answer from a node, you can't do it with topics. The services give us the possibility to interact with nodes. Also, services must have a unique name. When a node has a service, all the nodes can communicate with it, thanks to ROS client libraries.
- **Bags**: Bags are a format to save and play back the ROS message data. Bags are an important mechanism for storing data, such as sensor data, that can be difficult to collect but is necessary for developing and testing algorithms. You will use bags a lot while working with complex robots.

In the following figure, you can see the graphical representation of this level. It represents a real robot working in real conditions. In the graph, you can see the nodes, the topics, what node is subscribed to a topic, and so on. This graph does not represent the messages, bags, Parameter Server, and services. It is necessary for other tools to see a graphic representation of them. The tool used to create the graph is `rxgraph`; you will learn more about it in the upcoming sections.

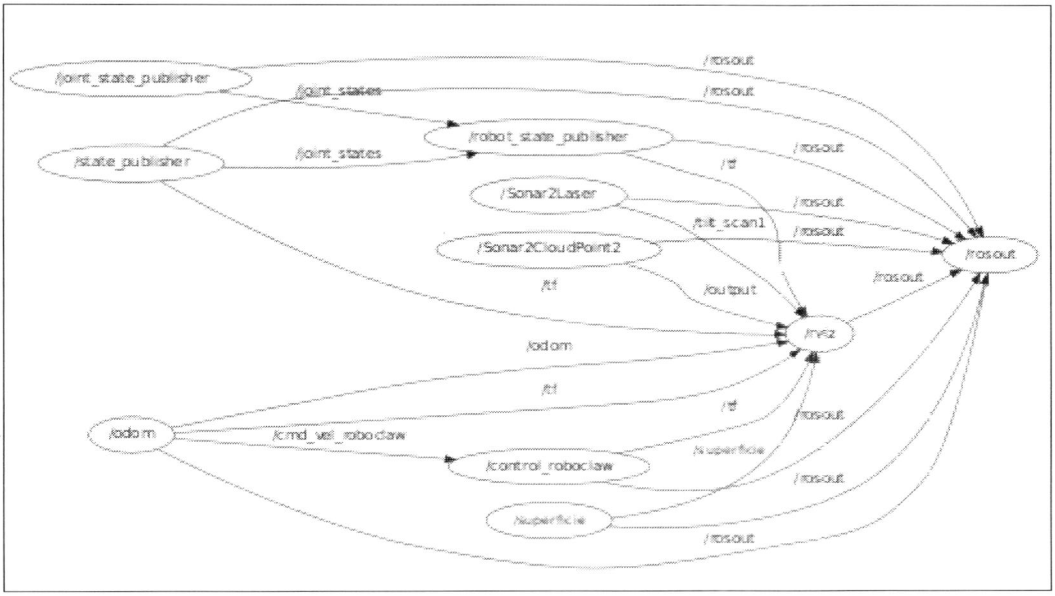

Nodes

Nodes are executables that can communicate with other processes using topics, services, or the Parameter Server. Using nodes in ROS provides us with fault tolerance and separates the code and functionalities making the system simpler.

A node must have a unique name in the system. This name is used to permit the node to communicate with another node using its name without ambiguity. A node can be written using different libraries such as `roscpp` and `rospy`; `roscpp` is for C++ and `rospy` is for Python. Throughout this book, we will be using `roscpp`.

ROS has tools to handle nodes and give us information about it such as `rosnode`. The tool `rosnode` is a command-line tool for displaying information about nodes, such as listing the currently running nodes. The commands supported are as follows:

- `rosnode info node`: This prints information about the node
- `rosnode kill node`: This kills a running node or sends a given signal
- `rosnode list`: This lists the active nodes
- `rosnode machine hostname`: This lists the nodes running on a particular machine or lists the machines
- `rosnode ping node`: This tests the connectivity to the node
- `rosnode cleanup`: This purges registration information from unreachable nodes

In the upcoming sections, we will learn how to use these commands with some examples.

A powerful feature of ROS nodes is the possibility to change parameters while you are starting the node. This feature gives us the power to change the node name, topic names, and parameter names. We use this to reconfigure the node without recompiling the code so that we can use the node in different scenes.

An example for changing a topic name is as follows:

```
$ rosrun book_tutorials tutorialX topic1:=/level1/topic1
```

This command will change the topic name `topic1` to `/level1/topic1`. I am sure you do not understand this at this moment, but you will find the utility of it in the upcoming chapters.

To change parameters in the node, you can do something similar to changing the topic name. For this, you only need to add an underscore to the parameter name; for example:

```
$ rosrun book_tutorials tutorialX _param:=9.0
```

This will set `param` to the float number `9.0`.

Keep in mind that you cannot use some names that are reserved by the system. They are:

- `__name`: This is a special reserved keyword for the name of the node
- `__log`: This is a reserved keyword that designates the location where the node's log file should be written
- `__ip` and `__hostname`: These are substitutes for `ROS_IP` and `ROS_HOSTNAME`
- `__master`: This is a substitute for `ROS_MASTER_URI`
- `__ns`: This is a substitute for `ROS_NAMESPACE`

Topics

Topics are buses used by nodes to transmit data. Topics can be transmitted without a direct connection between nodes, meaning the production and consumption of data are decoupled. A topic can have various subscribers.

Each topic is strongly typed by the ROS message type used to publish it, and nodes can only receive messages from a matching type. A node can subscribe to a topic only if it has the same message type.

The topics in ROS can be transmitted using TCP/IP and UDP. The TCP/IP-based transport is known as **TCPROS** and uses the persistent TCP/IP connection. This is the default transport used in ROS.

The UDP-based transport is known as **UDPROS** and is a low-latency, lossy transport. So, it is best suited for tasks such as teleoperation.

ROS has a tool to work with topics called `rostopic`. It is a command-line tool that gives us information about the topic or publishes data directly on the network.

This tool has the following parameters:

- `rostopic bw /topic`: This displays the bandwidth used by the topic.
- `rostopic echo /topic`: This prints messages to the screen.

- `rostopic find message_type`: This finds topics by their type.
- `rostopic hz /topic`: This displays the publishing rate of the topic.
- `rostopic info /topic`: This prints information about the active topic, the topics published, the ones it is subscribed to, and services.
- `rostopic list`: This prints information about active topics.
- `rostopic pub /topic type args`: This publishes data to the topic. It allows us to create and publish data in whatever topic we want, directly from the command line.
- `rostopic type /topic`: This prints the topic type, that is, the type of message it publishes.

We will learn to use it in the upcoming sections.

Services

When you need to communicate with nodes and receive a reply, you cannot do it with topics; you need to do it with services.

The services are developed by the user, and standard services don't exist for nodes. The files with the source code of the messages are stored in the `srv` folder.

Like topics, services have an associated service type that is the package resource name of the `.srv` file. As with other ROS filesystem-based types, the service type is the package name and the name of the `.srv` file. For example, the `chapter2_tutorials/srv/chapter2_srv1.srv` file has the service type `chapter2_tutorials/chapter2_srv1`.

ROS has two command-line tools to work with services, `rossrv` and `rosservice`. With `rossrv`, we can see information about the services' data structure, and it has the exact same usage as `rosmsg`.

With `rosservice`, we can list and query services. The commands supported are as follows:

- `rosservice call /service args`: This calls the service with the provided arguments
- `rosservice find msg-type`: This finds services by the service type
- `rosservice info /service`: This prints information about the service
- `rosservice list`: This lists the active services
- `rosservice type /service`: This prints the service type
- `rosservice uri /service`: This prints the service ROSRPC URI

Messages

A node sends information to another node using messages that are published by topics. The message has a simple structure that uses standard types or types developed by the user.

Message types use the following standard ROS naming convention: the name of the package, followed by `/`, and the name of the `.msg` file. For example, `std_msgs/msg/String.msg` has the message type, `std_msgs/String`.

ROS has the command-line tool `rosmsg` to get information about messages.

The accepted parameters are as follows:

- `rosmsg show`: This displays the fields of a message
- `rosmsg list`: This lists all the messages
- `rosmsg package`: This lists all the messages in a package
- `rosmsg packages`: This lists all packages that have the message
- `rosmsg users`: This searches for code files that use the message type
- `rosmsg md5`: This displays the MD5 sum of a message

Bags

A bag is a file created by ROS with the `.bag` format to save all the information of the messages, topics, services, and others. You can use this data later to visualize what has happened; you can play, stop, rewind, and perform other operations.

The bag file can be reproduced in ROS like a real session, sending the topics at the same time with the same data. Normally, we use this functionality to debug our algorithms.

To use bag files, we have the following tools in ROS:

- `rosbag`: This is used to record, play, and perform other operations
- `rxbag`: This is used to visualize the data in a graphic environment
- `rostopic`: This helps us see the topics sent to the nodes

Master

The ROS Master provides naming and registration services to the rest of the nodes in the ROS system. It tracks publishers and subscribers to topics as well as services. The role of the Master is to enable individual ROS nodes to locate one another. Once these nodes have located each other, they communicate with each other in a peer-to-peer fashion.

The Master also provides the Parameter Server. The Master is most commonly run using the `roscore` command, which loads the ROS Master along with other essential components.

Parameter Server

A Parameter Server is a shared, multivariable dictionary that is accessible via a network. Nodes use this server to store and retrieve parameters at runtime.

The Parameter Server is implemented using XML-RPC and runs inside the ROS Master, which means that its API is accessible via normal XMLRPC libraries.

The Parameter Server uses XMLRPC data types for parameter values, which include the following:

- 32-bit integers
- Booleans
- Strings
- Doubles
- ISO 8601 dates
- Lists
- Base 64-encoded binary data

ROS has the `rosparam` tool to work with the Parameter Server. The supported parameters are as follows:

- `rosparam list`: This lists all the parameters in the server
- `rosparam get parameter`: This gets the value of a parameter
- `rosparam set parameter value`: This sets the value of a parameter
- `rosparam delete parameter`: This deletes a parameter
- `rosparam dump file`: This saves the Parameter Server to a file
- `rosparam load file`: This loads a file (with parameters) on the Parameter Server

Understanding the ROS Community level

The ROS Community level concepts are ROS resources that enable separate communities to exchange software and knowledge. These resources include:

- **Distributions**: ROS distributions are collections of versioned stacks that you can install. ROS distributions play a similar role to Linux distributions. They make it easier to install a collection of software, and they also maintain consistent versions across a set of software.
- **Repositories**: ROS relies on a federated network of code repositories, where different institutions can develop and release their own robot software components.
- **The ROS Wiki**: The ROS Wiki is the main forum for documenting information about ROS. Anyone can sign up for an account and contribute their own documentation, provide corrections or updates, write tutorials, and more.
- **Mailing lists**: The ROS user-mailing list is the primary communication channel about new updates to ROS as well as a forum to ask questions about the ROS software.

Some tutorials to practice with ROS

It is time to practice what we have learned until now. In the upcoming sections, you will see examples to practice along with the creation of packages, using nodes, using the Parameter Server, and moving a simulated robot with TurtleSim.

Navigating through the ROS filesystem

We have some command-line tools to navigate through the filesystem. We are going to explain the most used ones.

To get information and move to packages and stacks, we will use `rospack`, `rosstack`, `roscd`, and `rosls`.

We use `rospack` and `rosstack` to get information about packages and stacks, the path, the dependencies, and so on.

For example, if you want to find the path of the `turtlesim` package, you will use this:

```
$ rospack find turtlesim
```

You will then obtain the following:

```
/opt/ros/fuerte/share/turtlesim
```

The same happens with stacks that you have installed in the system. An example of this is as follows:

```
$ rosstack find 'nameofstack'
```

To list the files inside the pack or stack, you will use this:

```
$ rosls turtlesim
```

You will then obtain the following:

```
cmake    images    mimic  srv         turtle_teleop_key
draw_square  manifest.xml  msg turtlesim_node
```

If you want to go inside the folder, you will use `roscd` as follows:

```
$ roscd turtlesim
$ pwd
```

You will obtain the following new path:

```
/opt/ros/fuerte/share/turtlesim
```

Creating our own workspace

Before doing anything, we are going to create our own workspace. In this workspace, we will have all the code that we will use in this book.

To see the workspace that ROS is using, use the following command:

```
$ echo $ROS_PACKAGE_PATH
```

You will see something like this:

```
/opt/ros/fuerte/share:/opt/ros/fuerte/stacks
```

The folder that we are going to create is in `~/dev/rosbook/`. To add this folder, we use the following lines:

```
$ cd ~
$ mkdir -p dev/rosbook
```

Once we have the folder in place, it is necessary to add this new path to `ROS_PACKAGE_PATH`. To do that, we only need to add a new line at the end of the `~/.bashrc` file:

```
$ echo "export ROS_PACKAGE_PATH"~/dev/rosbook:${ROS_PACKAGE_PATH}" >>
~/.bashrc
$ . ~/.bashrc
```

Now we have our new folder created and configured for use with ROS.

Creating an ROS package

As said before, you can create a package manually. To avoid tedious work, we will use the `roscreate-pkg` command-line tool.

We will create the new package in the folder created previously using the following lines:

```
$ cd ~/dev/rosbook
$ roscreate-pkg chapter2_tutorials std_msgs rospy roscpp
```

The format of this command includes the name of the package and the dependencies that will have the package; in our case, they are `std_msgs`, `rospy`, and `roscpp`. This is shown in the following command line:

```
roscreate-pkg [package_name] [depend1] [depend2] [depend3]
```

The dependencies included are the following:

- `std_msgs`: This contains common message types representing primitive data types and other basic message constructs, such as multiarrays.
- `rospy`: This is a pure Python client library for ROS
- `roscpp`: This is a C++ implementation of ROS. It provides a client library that enables C++ programmers to quickly interface with ROS topics, services, and parameters.

If everything is right, you will see the following screenshot:

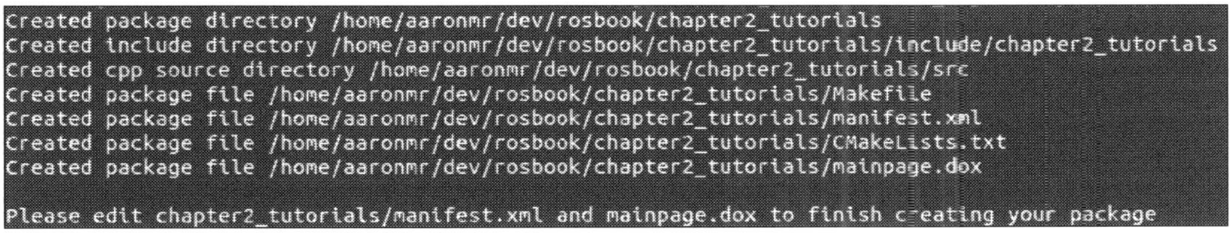

As we saw before, you can use the `rospack`, `roscd`, and `rosls` commands to get information of the new package:

- `rospack find chapter2_tutorials`: This command helps us find the path
- `rospack depends chapter2_tutorials`: This command helps us see the dependencies
- `rosls chapter2_tutorials`: This command helps us see the content
- `roscd chapter2_tutorials`: This command changes the actual path

Building an ROS package

Once you have your package created, and you have some code, it is necessary to build the package. When you build the package, what really happens is that the code is compiled.

To build a package, we will use the `rosmake` tool as follows:

```
$ rosmake chapter2_tutorials
```

After a few seconds, you will see something like this:

```
[ rosmake ] Results:
[ rosmake ] Cleaned 5 packages.
[ rosmake ] Built 5 packages with 0 failures.
[ rosmake ] Summary output to directory
[ rosmake ] /home/aaronmr/.ros/rosmake/rosmake_output-20130713-183756
```

If you don't obtain failures, the package is compiled.

Playing with ROS nodes

As we have explained in the *Nodes* section, nodes are executable programs and these executables are in the `packagename/bin` directory. To practice and learn about nodes, we are going to use a typical package called `turtlesim`.

If you have installed the desktop installation file, you have the `turtlesim` package installed; if not, install it using the following command:

```
$ sudo apt-get install ros-fuerte-ros-tutorials
```

Before starting with anything, you must start `roscore` as follows:

```
$ roscore
```

To get information on nodes, we have the tool `rosnode`. To see what parameters are accepted, type the following command:

```
$ rosnode
```

You will obtain a list of accepted parameters as shown in the following screenshot:

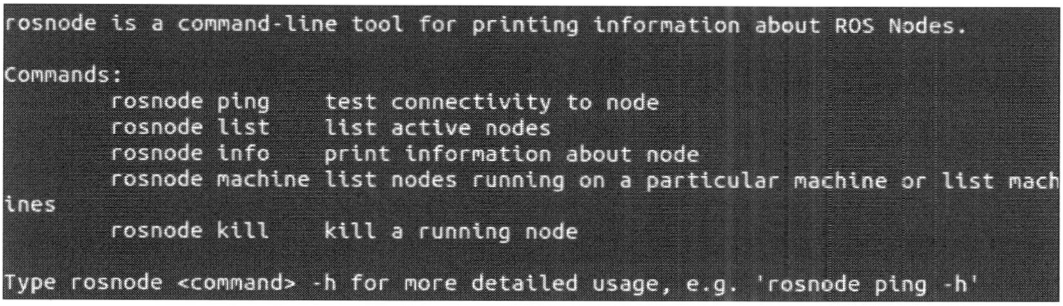

If you want a more detailed explanation on the use of these parameters, use the following line:

```
$ rosnode <param> -h
```

Now that `roscore` is running, we are going to get information about the nodes that are running:

```
$ rosnode list
```

You see that the only node running is `/rosout`. It is normal because this node runs whenever `roscore` is run.

We can get all the information of this node using parameters. Try using the following commands for more information:

```
$ rosnode info
$ rosnode ping
$ rosnode machine
$ rosnode kill
```

Now we are going to start a new node with `rosrun` as follows:

```
$ rosrun turtlesim turtlesim_node
```

We will then see a new window appear with a little turtle in the middle, as shown in the following screenshot:

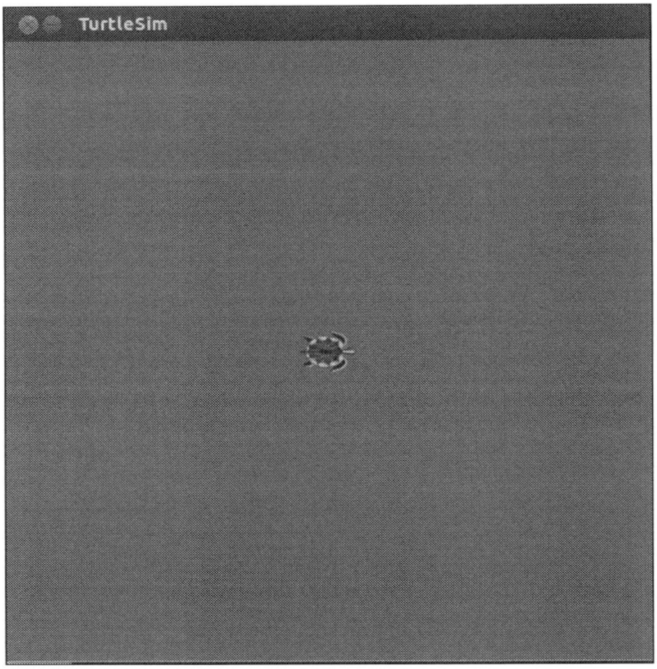

If we see the node list now, we will see a new node with the name `/turtlesim`.

You can see the information of the node using `rosnode info nameNode`. You can see a lot of information that can be used to debug your programs:

```
$ rosnode info /turtlesim
Node [/turtlesim]
Publications:
 * /turtle1/color_sensor [turtlesim/Color]
 * /rosout [rosgraph_msgs/Log]
 * /turtle1/pose [turtlesim/Pose]
Subscriptions:
 * /turtle1/command_velocity [unknown type]
Services:
 * /turtle1/teleport_absolute
 * /turtlesim/get_loggers
```

```
 * /turtlesim/set_logger_level
 * /reset
 * /spawn
 * /clear
 * /turtle1/set_pen
 * /turtle1/teleport_relative
 * /kill
contacting node http://aaronmr-laptop:42791/ ...
Pid: 28337
Connections:
 * topic: /rosout
  * to: /rosout
  * direction: outbound
  * transport: TCPROS
```

In the information, we can see the publications (topics), subscriptions (topics), and the services (srv) that the node has and the unique names of each.

Now, let us show you how to interact with the node using topics and services.

Learning how to interact with topics

To interact and get information of topics, we have the `rostopic` tool. This tool accepts the following parameters:

- `rostopic bw`: This displays the bandwidth used by topics
- `rostopic echo`: This prints messages to the screen
- `rostopic find`: This finds topics by their type
- `rostopic hz`: This displays the publishing rate of topics
- `rostopic info`: This prints information about active topics
- `rostopic list`: This lists the active topics
- `rostopic pub`: This publishes data to the topic
- `rostopic type`: This prints the topic type

If you want to see more information on these parameters, use `-h` as follows:

```
$ rostopic bw -h
```

With the `pub` parameter, we can publish topics that can subscribe to any node. We only need to publish the topic with the correct name. We will do this test later; we are now going to use a node that will do this work for us:

```
$ rosrun turtlesim turtle_teleop_key
```

With this node, we can move the turtle using the arrow keys as you can see in the following screenshot:

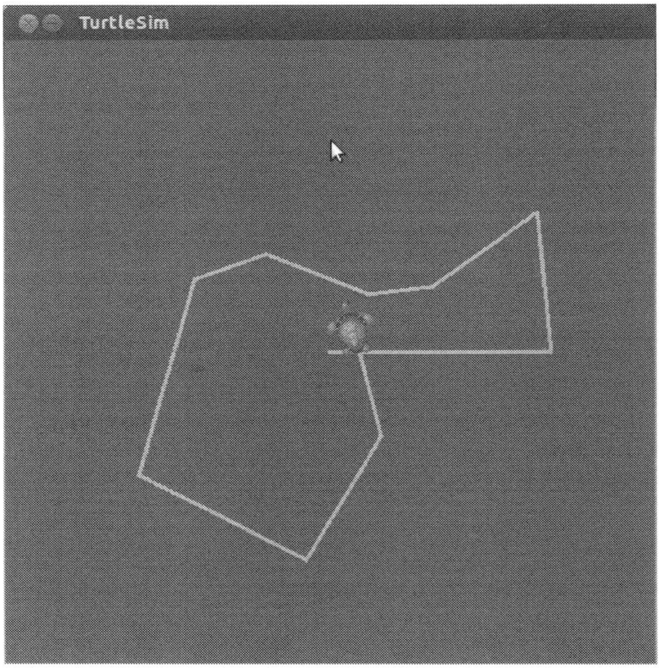

Why is the turtle moving when `turtle_teleop_key` is executed?

If you want to see the information of the `/teleop_turtle` and `/turtlesim` nodes, we can see in the following code that there exists a topic called `* /turtle1/command_velocity [turtlesim/Velocity]` in the `Publications` section of the first node; in the `Subscriptions` section of the second node, there is `* /turtle1/command_velocity [turtlesim/Velocity]`:

```
$ rosnode info /teleop_turtle

Node [/teleop_turtle]
...
```

```
Publications:
 * /turtle1/command_velocity [turtlesim/Velocity]
 ...

$ rosnode info /turtlesim

Node [/teleop_turtle]
...
Subscriptions:
 * /turtle1/command_velocity [turtlesim/Velocity]
...
```

This means that the first node is publishing a topic that the second node can subscribe to.

You can see the topics' list using the following command lines:

```
$ rostopic list

/rosout
/rosout_agg
/turtle1/color_sensor
/turtle1/command_velocity
/turtle1/pose
```

With the `echo` parameter, you can see the information sent by the node.

Run the following command line and use the arrow keys to see what data is being sent:

```
$ rostopic echo /turtle1/command_velocity
```

You will see something like this:

```
linear: 2.0
angular: 0.0
---
linear: 0.0
angular: 2.0
---
```

You can see the type of message sent by the topic using the following command lines:

```
$  rostopic type /turtle1/command_velocity


turtlesim/Velocity
```

If you want to see the message fields, you can do it with the following command lines:

```
$ rosmsg show turtlesim/Velocity


float32 linear
float32 angular
```

These tools are useful because, with this information, we can publish topics using the command, `rostopic pub [topic] [msg_type] [args]`:

```
$ rostopic pub -1 /turtle1/command_velocity turtlesim/Velocity -- 1 1
```

You will see the turtle doing a curve.

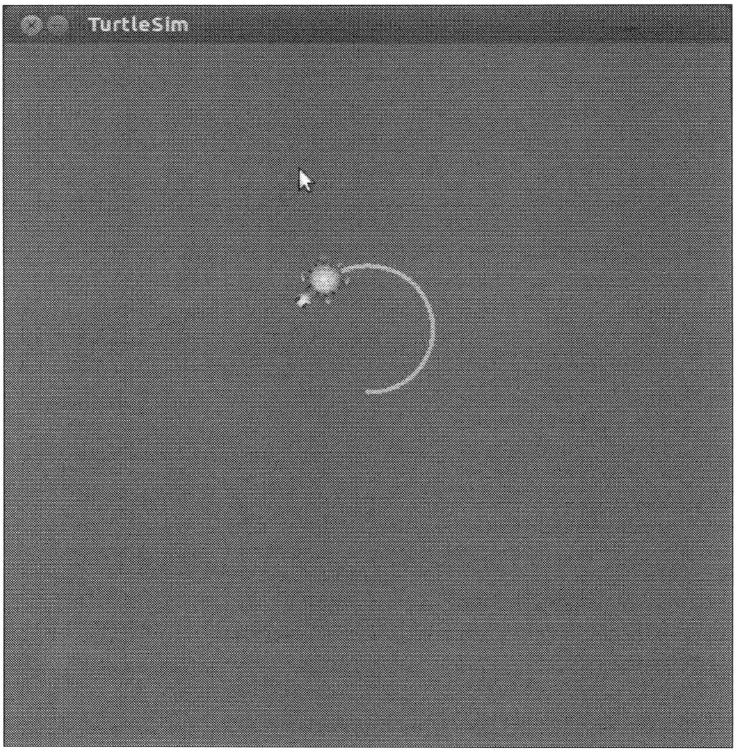

Learning how to use services

Services are another way through which nodes can communicate with each other. Services allow nodes to send a request and receive a response.

The tool that we are going to use to interact with services is `rosservice`. The accepted parameters for this command are as follows:

- `rosservice args /service`: This prints service arguments
- `rosservice call /service`: This calls the service with the provided arguments
- `rosservice find msg-type`: This finds services by their service type
- `rosservice info /service`: This prints information about the service
- `rosservice list`: This lists active services
- `rosservice type /service`: This prints the service type
- `rosservice uri /service`: This prints the service ROSRPC URI

We are going to list the services available for the `turtlesim` node by using the following code, so if it is not working, run `roscore` and run the `turtlesim` node:

```
$ rosservice list
```

You will obtain the following output:

```
/clear
/kill
/reset
/rosout/get_loggers
/rosout/set_logger_level
/spawn
/teleop_turtle/get_loggers
/teleop_turtle/set_logger_level
/turtle1/set_pen
/turtle1/teleport_absolute
/turtle1/teleport_relative
/turtle2/set_pen
/turtle2/teleport_absolute
/turtle2/teleport_relative
/turtlesim/get_loggers
/turtlesim/set_logger_level
```

If you want to see the type of any service, for example, the `/clear` service, use:

```
$ rosservice type /clear
```

You will then obtain:

```
std_srvs/Empty
```

To invoke a service, you will use `rosservice call [service] [args]`. If you want to invoke the `/clear` service, use:

```
$ rosservice call /clear
```

In the **turtlesim** window, you will now see that the lines created by the movements of the turtle will be deleted.

Now, we are going to try another service, for example, the `/spawn` service. This service will create another turtle in another location with a different orientation. To start with, we are going to see the following type of message:

```
$ rosservice type /spawn | rossrv show
```

We will then obtain the following:

```
float32 x
float32 y
float32 theta
string name
---
string name
```

With these fields, we know how to invoke the service. We need the positions of x and y, the orientation (theta), and the name of the new turtle:

```
$ rosservice call 3 3 0.2 "new_turtle"
```

We then obtain the following result:

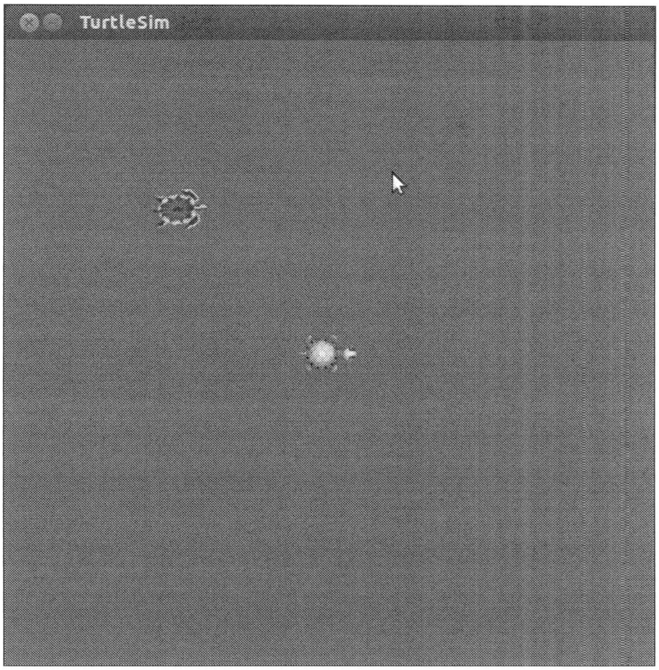

Using the Parameter Server

The Parameter Server is used to store data that is accessible by all the nodes. ROS has a tool to manage the Parameter Server called `rosparam`. The accepted parameters are as follows:

- `rosparam set parameter value`: This sets the parameter
- `rosparam get parameter`: This gets the parameter
- `rosparam load file`: This loads parameters from the file
- `rosparam dump file`: This dumps parameters to the file
- `rosparam delete parameter`: This deletes the parameter
- `rosparam list`: This lists the parameter names

For example, we can see the parameters in the server that are used by all the nodes:

```
$ rosparam list
```

We obtain the following output:

```
/background_b
/background_g
/background_r
/rosdistro
/roslaunch/uris/host_aaronmr_laptop__60878
/rosversion
/run_id
```

The `background` parameters are of the `turtlesim` node. These parameters change the color of the windows that are initially blue. If you want to read a value, you will use the `get` parameter:

```
$ rosparam get /background_b
```

To set a new value, you will use the `set` parameter:

```
$ rosparam set /background_b 100
```

Another important feature of `rosparam` is the `dump` parameter. With this parameter, you can save or load the content of the Parameter Server.

To save the Parameter Server, use `rosparam dump [file_name]`:

```
$ rosparam dump save.yaml
```

To load a file with new data for the Parameter Server, use `rosparam load [file_name] [namespace]`:

```
$ rosparam load load.yaml namespace
```

Creating nodes

In this section, we are going to learn how to create two nodes: one to publish some data and the other to receive this data. This is the basic way of communicating between two nodes, that is, to handle data and to do something with that data:

```
$ roscd chapter2_tutorials/src/
```

Create two files with the names, `example1_a.cpp` and `example1_b.cpp`.

The `example1_a.cpp` file will send the data with the node name, and the `example2example1_b.cpp` file will show the data in the shell.

Copy the following code into the `example1_a.cpp` file or download it from the repository:

```
#include "ros/ros.h"
#include "std_msgs/String.h"
#include <sstream>

int main(int argc, char **argv)
{
  ros::init(argc, argv, "example1_a");
  ros::NodeHandle n;
  ros::Publisher chatter_pub = n.advertise<std_
msgs::String>("message", 1000);
  ros::Rate loop_rate(10);
  while (ros::ok())
  {
    std_msgs::String msg;
    std::stringstream ss;
    ss << " I am the example1_a node ";
    msg.data = ss.str();
    //ROS_INFO("%s", msg.data.c_str());
    chatter_pub.publish(msg);
    ros::spinOnce();
    loop_rate.sleep();
  }
  return 0;
}
```

Here is some further explanation about the preceding code. The headers to be included are `ros/ros.h`, `std_msgs/String.h`, and `sstream`. Here, `ros/ros.h` includes all the necessary files for using the node with ROS, and `std_msgs/String.h` includes the header that denotes the type of message we are going to use:

```
#include "ros/ros.h"
#include "std_msgs/String.h"
#include <sstream>
```

Initiate the node and set the name; remember that the name must be unique:

```
ros::init(argc, argv, "example1_a");
```

The following is the handler of our process:

```
ros::NodeHandle n;
```

Set a publisher and tell the Master the name of the topic and the type. The name is `message`, and the second parameter is the buffer size. If the topic is publishing data quickly, the buffer will keep 1,000 messages as follows:

```
ros::Publisher chatter_pub = n.advertise<std_msgs::String>("message",
1000);
```

Set the frequency to send the data; in this case, it is 10 Hz:

```
ros::Rate loop_rate(10);
```

The `ros::ok()` library stops the node if it receives a *Ctrl* + *C* keypress or if ROS stops all the nodes:

```
while (ros::ok())
{
```

In this part, we create a variable for the message with the correct type to send the data:

```
std_msgs::String msg;
std::stringstream ss;
ss << " I am the example1_a node ";
msg.data = ss.str();
```

Here, the message is published:

```
chatter_pub.publish(msg);
```

We have a subscriber in this part where ROS updates and reads all the topics:

```
ros::spinOnce();
```

Sleep the necessary time to get a 10 Hz frequency:

```
loop_rate.sleep();
```

Copy the following code into the `example1_b.cpp` file or download it from the repository:

```
#include "ros/ros.h"
#include "std_msgs/String.h"

void chatterCallback(const std_msgs::String::ConstPtr& msg)
{
  ROS_INFO("I heard: [%s]", msg->data.c_str());
}

int main(int argc, char **argv)
```

```
{
  ros::init(argc, argv, "example1_b");
  ros::NodeHandle n;
  ros::Subscriber sub = n.subscribe("message", 1000, chatterCallback);
  ros::spin();
  return 0;
}
```

Here is some further explanation about the preceding code. Include the headers and the type of message to use for the topic:

```
#include "ros/ros.h"
#include "std_msgs/String.h"
```

This function is called every time that node receives a message. This is where we do something with the data; in this case, we show it in the shell:

`ROS_INFO()` is used to print it in the shell.

```
void messageCallback(const std_msgs::String::ConstPtr& msg)
{
  ROS_INFO("I heard: [%s]", msg->data.c_str());
}
```

Create a subscriber and start to listen to the topic with the name `message`. The buffer will be of 1,000, and the function to handle the message will be `messageCallback`:

```
ros::Subscriber sub = n.subscribe("message", 1000, messageCallback);
```

The `ros::spin()` library is a loop where the node starts to read the topic, and when a message arrives, `messageCallback` is called. When the user presses *Ctrl* + *C*, the node will exit the loop and the loop ends:

```
ros::spin();
```

Building the node

As we are using the package `chapter2_tutorials`, we are going to edit the file `CMakeLists.txt`. You can use your favorite editor or use the `rosed` tool. This will open the file with the Vim editor:

```
$ rosed chapter2_tutorials CMakeLists.txt
```

At the end of the file, we copy the following command lines:

```
rosbuild_add_executable(example1_a src/example1_a.cpp)
rosbuild_add_executable(example1_b src/example1_b.cpp)
```

These lines will create two executables in the `/bin` folder.

Now to build the package and compile all the nodes, use the `rosmake` tool:

```
$ rosmake chapter2_tutorials
```

If ROS is not running on your computer, you will have to use:

```
$ roscore
```

You can check whether ROS is running using the `rosnode list` command as follows:

```
$ rosnode list
```

Now, run both the nodes in different shells:

```
$ rosrun chapter2_tutorials example1_a
$ rosrun chapter2_tutorials example1_b
```

If you check the shell where the node `example1_b` is running, you will see something like this:

```
...
[ INFO] [1355228542.283236850]: I heard: [ I am the example1_a node ]
[ INFO] [1355228542.383221843]: I heard: [ I am the example1_a node ]
[ INFO] [1355228542.483249861]: I heard: [ I am the example1_a node ]
...
```

Everything that is happening can be viewed in the following figure. You can see that the `example1_a` node is publishing the topic `message`, and the `example2_b` node is subscribing to the topic.

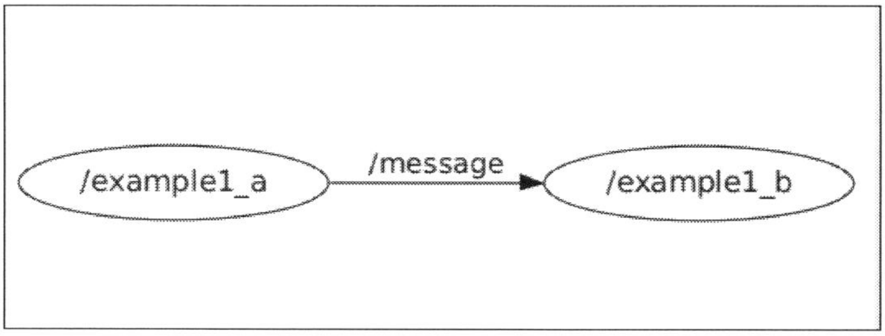

You can use `rosnode` and `rostopic` to debug and see what the nodes are doing.

Try the following commands:

```
$ rosnode list
$ rosnode info /example1_a
```

```
$ rosnode info /example1_b
$ rostopic list
$ rostopic info /message
$ rostopic type /message
$ rostopic bw /message
```

Creating msg and srv files

In this section, we are going to learn how to create `msg` and `srv` files for using them in our nodes. They are files where we put a specification about the type of data to be transmitted and the values of these data. ROS will use these files to create the necessary code for us to implement the `msg` and `srv` files to be used in our nodes. Let's start with the `msg` file first.

In the example used in the *Building the node* section, we have created two nodes with a standard type `message`. Now, we are going to learn how to create custom messages with the tools that ROS has.

First, create a new folder `msg` in our package `chapter2_tutorials`, and create a new file `chapter2_msg1` adding the following lines:

```
int32 A
int32 B
int32 C
```

Now edit `CMakeList.txt`, remove `#` from the line `# rosbuild_genmsg()`, and build the package with `rosmake` as follows:

```
$ rosmake chapter2_tutorials
```

To check whether all is OK, you can use the `rosmsg` command:

```
$ rosmsg show chapter2_tutorials/chapter2_msg1
```

If you see the same content of the file `chapter2_msg1.msg`, all is OK.

Now we are going to create an `srv` file. Create a new folder in the `chapter2_tutorials` folder with the name `srv` and create a new file `chapter2_srv1.srv` adding the following lines:

```
int32 A
int32 B
int32 C
---
int32 sum
```

Edit `CMakeList.txt`, and remove # from the line `# rosbuild_gensrv()`, and build the package with `rosmake chapter2_tutorials`.

You can test whether all is OK using the `rossrv` tool as follows:

```
$ rossrv show chapter2_tutorials/chapter2_srv1
```

If you see the same content of the file `chapter2_srv1.srv`, all is OK.

Using the new srv and msg files

First, we are going to learn how to create a service and how to use it in ROS. Our service will calculate the sum of three numbers. We need two nodes, a server and a client.

In the package `chapter2_tutorials`, create two new nodes with names: `example2_a.cpp` and `example2_b.cpp`. Remember to put the files in the `src` folder.

In the first file, `example2_a.cpp`, add this code:

```
#include "ros/ros.h"
#include "chapter2_tutorials/chapter2_srv1.h"

bool add(chapter2_tutorials::chapter2_srv1::Request  &req,
         chapter2_tutorials::chapter2_srv1::Response &res)
{
  res.sum = req.A + req.B + req.C;
  ROS_INFO("request: A=%ld, B=%ld C=%ld", (int)req.A, (int)req.B,
(int)req.C);
  ROS_INFO("sending back response: [%ld]", (int)res.sum);
  return true;
}

int main(int argc, char **argv)
{
  ros::init(argc, argv, "add_3_ints_server");
  ros::NodeHandle n;

  ros::ServiceServer service = n.advertiseService("add_3_ints", add);
  ROS_INFO("Ready to add 3 ints.");
  ros::spin();

  return 0;
}
```

Include the necessary headers and `srv` that we created:

```
#include "ros/ros.h"
#include "chapter2_tutorials/chapter2_srv1.h"
```

This function will add three variables and send the result to the other node:

```
bool add(chapter2_tutorials::chapter2_srv1::Request  &req,
         chapter2_tutorials::chapter2_srv1::Response &res)
```

Here, the service is created and advertised over ROS as follows:

```
ros::ServiceServer service = n.advertiseService("add_3_ints", add);
```

In the second file, `example2_b.cpp`, add this code:

```
#include "ros/ros.h"
#include "chapter2_tutorials/chapter2_srv1.h"
#include <cstdlib>

int main(int argc, char **argv)
{
  ros::init(argc, argv, "add_3_ints_client");
  if (argc != 4)
  {
    ROS_INFO("usage: add_3_ints_client A B C ");
    return 1;
  }

  ros::NodeHandle n;
  ros::ServiceClient client = n.serviceClient<chapter2_
tutorials::chapter2_srv1>("add_3_ints");
  chapter2_tutorials::chapter2_srv1 srv;
  srv.request.A = atoll(argv[1]);
  srv.request.B = atoll(argv[2]);
  srv.request.C = atoll(argv[3]);
  if (client.call(srv))
  {
    ROS_INFO("Sum: %ld", (long int)srv.response.sum);
  }
  else
  {
    ROS_ERROR("Failed to call service add_3_ints");
    return 1;
  }

  return 0;
}
```

Create a client for the service with the name `add_3_ints`:

```
ros::ServiceClient client = n.serviceClient<chapter2_
tutorials::chapter2_srv1>("add_3_ints");
```

Here we create an instance of our `srv` file and fill all the values to be sent. If you remember, the message has three fields:

```
chapter2_tutorials::chapter2_srv1 srv;
srv.request.A = atoll(argv[1]);
srv.request.B = atoll(argv[2]);
srv.request.C = atoll(argv[3]);
```

With these lines, the service is called and the data is sent. If the call succeeds, `call()` will return `true`; if not, `call()` will return `false`:

```
if (client.call(srv))
```

To build new nodes, edit `CMakeList.txt` and add the following lines:

```
rosbuild_add_executable(example2_a src/example2_a.cpp)
rosbuild_add_executable(example2_b src/example2_b.cpp)
```

Now execute the following command:

```
$ rosmake chapter2_tutorials
```

To start the nodes, execute the following command lines:

```
$rosrun chapter2_tutorials example2_a
$rosrun chapter2_tutorials example2_b 1 2 3
```

And you should see something like this:

```
Node example2_a
[ INFO] [1355256113.014539262]: Ready to add 3 ints.
[ INFO] [1355256115.792442091]: request: A=1, B=2 C=3
[ INFO] [1355256115.792607196]: sending back response: [6]


Node example2_b
[ INFO] [1355256115.794134975]: Sum: 6
```

Now we are going to create nodes with our custom `msg` file. The example is the same, that is, `example1_a.cpp` and `example1_a.cpp` but with the new message `chapter2_msg1.msg`.

The following code snippet is present in the `example3_a.cpp` file:

```
#include "ros/ros.h"
#include "chapter2_tutorials/chapter2_msg1.h"
#include <sstream>

int main(int argc, char **argv)
{
  ros::init(argc, argv, "example1_a");
  ros::NodeHandle n;
  ros::Publisher pub = n.advertise<chapter2_tutorials::chapter2_
msg1>("message", 1000);
  ros::Rate loop_rate(10);
  while (ros::ok())
  {
    chapter2_tutorials::chapter2_msg1 msg;
    msg.A = 1;
    msg.B = 2;
    msg.C = 3;
    pub.publish(msg);
    ros::spinOnce();
    loop_rate.sleep();
  }
  return 0;
}
```

The following code snippet is present in the `example3_b.cpp` file:

```
#include "ros/ros.h"
#include "chapter2_tutorials/chapter2_msg1.h"

void messageCallback(const chapter2_tutorials::chapter2_
msg1::ConstPtr& msg)
{
  ROS_INFO("I heard: [%d] [%d] [%d]", msg->A, msg->B, msg->C);
}

int main(int argc, char **argv)
{
  ros::init(argc, argv, "example1_b");
  ros::NodeHandle n;
  ros::Subscriber sub = n.subscribe("message", 1000, messageCallback);
  ros::spin();
  return 0;
}
```

If we run both the nodes now, we will see something like this:

```
...
[ INFO] [1355270835.920368620]: I heard: [1] [2] [3]
[ INFO] [1355270836.020326372]: I heard: [1] [2] [3]
[ INFO] [1355270836.120367449]: I heard: [1] [2] [3]
[ INFO] [1355270836.220266466]: I heard: [1] [2] [3]
...
```

Summary

This chapter provides you with general knowledge of the ROS architecture and how it works. You saw some concepts, tools, and samples of how to interact with nodes, topics, and services. In the beginning, all these concepts could look complicated and without use, but in the upcoming chapters, you will start to understand the applications of these.

It is useful to practice with these terms and tutorials before continuing because, in the upcoming chapters, we will suppose that you know all the concepts and uses.

Remember that if you have queries regarding something, and you cannot find the solution in this book, you can use the official resources of ROS from the URL `http://www.ros.org`. Additionally, you can ask questions to the ROS community at `http://answers.ros.org`.

In the next chapter, you will learn how to debug and visualize data using ROS tools. These will help you find problems and to know whether what ROS is doing is correct, and what you expect it to do.

3
Debugging and Visualization

The ROS framework comes with a great number of powerful tools to help the user and developer in the process of debugging the code, and detecting problems with both the hardware and software. This comprises debugging facilities such as log messages as well as visualization and inspection capabilities, which allows the user to see what is going on in the system easily.

Here, we also cover the workflow to debug ROS nodes using GDB debugger as an example. Although this is almost the same as debugging a regular C/C++ program, there are a few aspects that must be taken into account. We will only focus on these particular aspects, since explaining the way to use the debugger is far from the scope of this chapter. You are encouraged to read the GDB reference and user manual for this.

ROS provides an API for logging, which allows setting different logging levels, depending on the semantics of the message to output or print. This is not only with the aim of helping debugging but also to have more informative and robust programs in case of failure. As we will see later, we can inform the user about the stages in the process of an algorithm with high-level informative messages, while also warning the user about missed values or parameters as well as regular or fatal errors, which are unrecoverable.

However, once our program compiles and runs, it might still fail. At this point, two major things could be happening; first, a problem related to the nodes, topics, services, or any other ROS element, or second, an issue caused by our algorithm itself. ROS provides a set of powerful tools to inspect the state of the system, which include the node's graph, with all the connections (publishers and subscribers) among topics as shown in the following screenshot. Both local and remote nodes are seamlessly addressed, so the user can easily and rapidly detect a problem in a node that is not running or a missed topic connection.

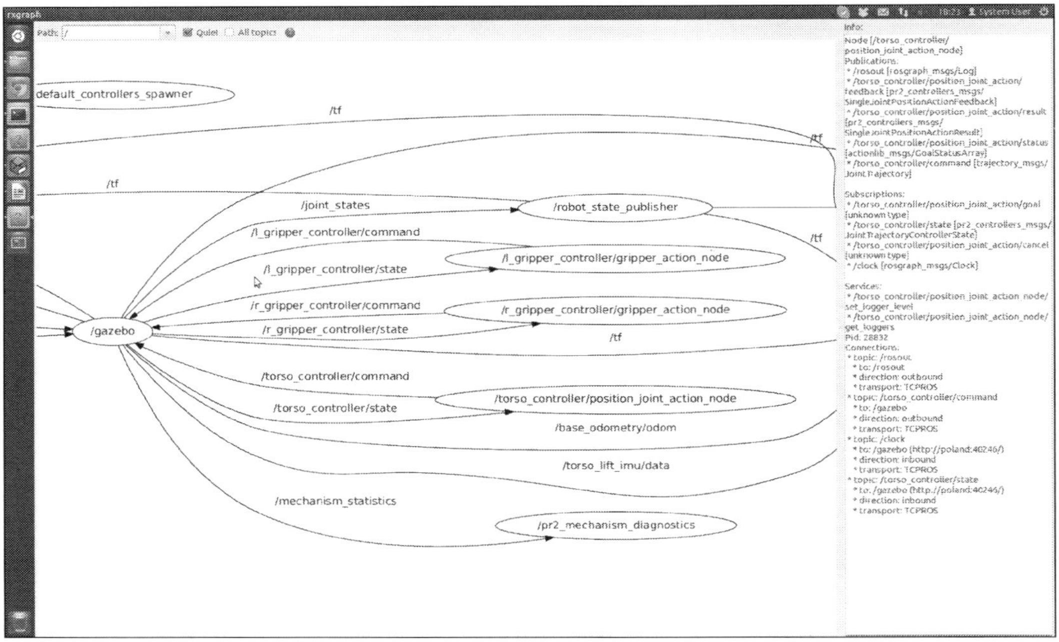

Up to some extent, a bunch of generic plotting tools are provided to analyze the output or results of our own algorithms so that it becomes easier to detect bugs. First of all, we have time series plots for scalar values, which might be fields of the messages transferred between nodes. Then, there are tools to show images, even with support for stereo pairs. Last but not least, we have 3D visualization tools such as rviz, as shown in the following screenshot, for the PR2 robot. They allow rendering point clouds, laser scans, and so on. As an important characteristic, all data belongs to a topic that is placed on a frame so that the data is rendered in that frame. A robot is generally a compound of many frames with transform frames among them. To help the user to see the relationships among them, we also have tools to watch the frame hierarchy at a glance.

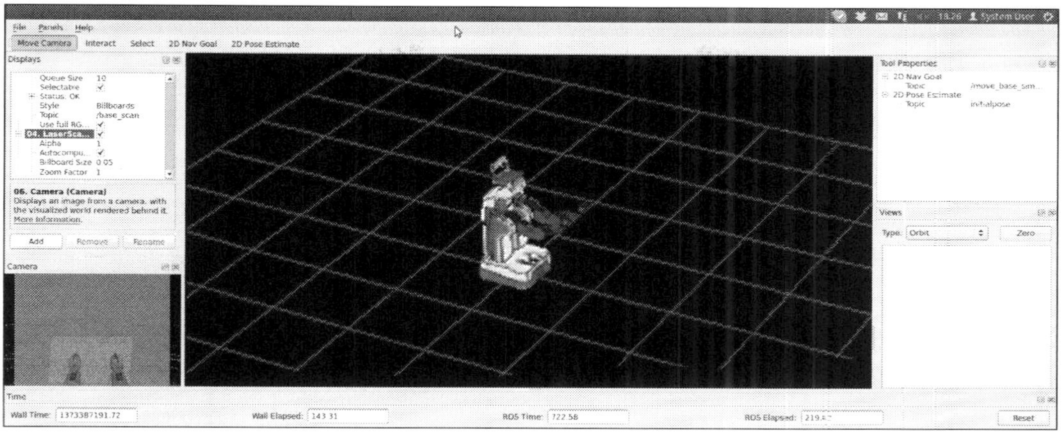

In the upcoming sections, we will cover the following aspects:

- Debugging and good practices for developing code when creating ROS nodes.
- Adding logging messages to our code and setting different levels, from debug messages to errors, or even fatal ones.
- Giving names, applying conditions, and throttling the logging messages, which becomes very useful in large projects.
- Presenting a graphical tool to manage all the messages.
- Inspecting the state of the ROS system by listing the nodes running and the topics and services available.
- Visualizing the node's graph representation, which are connected by publishing and subscribing to topics.
- Plotting scalar data of certain messages.
- Visualizing scalar data for complex types. In particular, we will cover images and the case of FireWire cameras, which are seamlessly supported in ROS, as well as the 3D visualization of many topic types.
- Explaining what frames are and their close relationship with the messages published by the topics. Similarly, we will see what a frame transformation in the TF tree is.
- Saving the messages sent by topics and how to replay them for simulation or evaluation purposes.

Debugging ROS nodes

In order to detect problems in the algorithms implemented inside ROS nodes, we can face the problem at different levels to make the debugging of the software easier. First, we must provide some readable information about the progress of the algorithm, driver, or another piece of software. In ROS, we have a set of logging macros for this particular purpose, which are completely integrated with the whole system. Second, we need tools to determine which verbosity level is desired for a given node; this is related to the ability to configure different logging levels. In the case of ROS, we will see how it is possible to set debugging/logging levels even on the fly as well as conditions and names for particular messages. Third, we must be able to use a debugger to step over the source code. We will see that the widely known GDB debugger integrates seamlessly with ROS nodes. Finally, at the abstraction (or semantic) level of ROS nodes and topics, it is useful to inspect the current state of the whole system. Such introspection capabilities in ROS are supported by means of tools that draw the nodes graph with connections among topics. The user or developer can easily detect a broken connection at a glance, as we will explain later in another section.

Using the GDB debugger with ROS nodes

We will start with the standard way of debugging C/C++ executables of any kind. The flexibility of ROS allows using the well-known GDB debugger with a regular C/C++ program. All we have to know is the location of our executable, which in the case of ROS would be a node implemented in C++. Therefore, we only have to move to the path that contains the node binary and run it within GDB. Indeed, we could also run our node directly without the typical rosrun <package> <node> syntax.

To make it simple, we will show you how to run a node in GDB for the `example1` node in the `chapter3_tutorials` package. First, move to the package with `roscd` as follows:

```
roscd chapter3_tutorials
```

Then, we only have to recall that C++ binaries are created inside the `bin` folder of the package folder's structure. Hence, we simply run it inside GDB using the following command:

```
gdb bin/example1
```

> Remember that you must have a `roscore` command running before you start your node because it will need the master/server running.

Once `roscore` is running, you can start your node in GDB by pressing the *R* key and *Enter*. You can also list the associated source code with the *L* key as well as set breakpoints or any of the functionalities that GDB comes with. If everything works correctly, you should see the following output in the GDB terminal after you have run the node inside the debugger:

```
(gdb) r
Starting program: /home/enrique/dev/rosbook/chapter3_tutorials/bin/
example1
[Thread debugging using libthread_db enabled]
Using host libthread_db library "/lib/x86_64-linux-gnu/libthread_
db.so.1".
[New Thread 0x7ffff2664700 (LWP 3204)]
...
...
[Thread 0x7ffff1e63700 (LWP 3205) exited]
[Inferior 1 (process 3200) exited normally]
```

Attaching a node to GDB while launching ROS

If we have a `launch` file to start our node, we have some attribute in the XML syntax that allows us to attach the node to a GDB session. For the previous node, `example1`, in the package `chapter3_tutorials`, we will add the following node element to the `launch` file:

```
<launch>
<node pkg="chapter3_tutorials" type="example1"
            name="example1"/>
</launch>
```

Note that the package is passed to the `pkg` attribute and the node to the `type` attribute. We also have to give this instance of the node a name since we can run more than one instance of the same node. In this case, we gave the same name as the node type, that is, the `name` attribute, which has the value `example1`. It is also a good practice to set the attribute `output` to `screen`, so the debugging messages, which we will see in the following code snippet, appear on the same terminal where we launched the node:

```
<node pkg="chapter3_tutorials" type="example1"
            name="example1" output="screen"/>
```

To attach it to GDB, we must add `launch-prefix="xterm -e gdb --args"`:

```
<launch>
<node pkg="chapter3_tutorials" type="example1"
            name="example1" output="screen"
            launch-prefix="xterm -e gdb --args"/>
</launch>
```

What this prefix does is very simple. It starts GDB, loads our node, and waits until the user presses the *C* or *R* key; that is, the node is loaded but waiting to run. This way the user can set breakpoints before the node runs and interact as a regular GDB session. Also, note that a new window opens up. This is because we create the GDB session in xterm, so we have a debugging window separated from the program output window.

Additionally, we can use the same attribute to attach the node to other diagnostic tools; for example, we can run `valgrind` on our program to detect memory leaks and perform some profiling analysis. For further information on `valgrind`, you can check out `http://valgrind.org`. To attach our node to it, we proceed in a similar way as we did with GDB. In this case, we do not need an additional window, so that we do not start xterm, and simply set `valgrind` as the launch prefix:

```
<launch>
<node pkg="chapter3_tutorials" type="example1"
            name="example1" output="screen"
            launch-prefix="valgrind"/>
</launch>
```

Enabling core dumps for ROS nodes

Although ROS nodes are actually regular executables, there are some tricky points to note to enable core dumps that can be used later in a GDB session. First of all, we have to set an unlimited core size. Note that this is required for any executable, not just ROS nodes:

```
ulimit -c unlimited
```

Then, to allow core dumps to be created, we must set the core filename to use the process `pid` by default; otherwise, they will not be created because at `$ROS_HOME`, there is already a core directory to prevent core dumps. Therefore, in order to create core dumps with the name and path `$ROS_HOME/core.PID`, we must do the following:

```
echo 1 > /proc/sys/kernel/core_uses_pid
```

Debugging messages

It is good practice to include messages that indicate what the program is doing. However, we must do it without compromising the efficiency of our software and the clearance of its output. In ROS, we have an API that covers both features and is built on top of log4cxx (a port of the well-known log4j logger library). In brief, we have several levels of messages, which might have a name depending on a condition or even throttle, with a null footprint on the performance and full integration with other tools in the ROS framework. Also, they are integrated seamlessly with the concurrent execution of nodes, that is, the messages do not get split, but they can be interleaved according to their timestamps. In the following sections, we will explain the details and how to use them adequately.

Outputting a debug message

ROS comes with a great number of functions or macros that allow us to output a debugging message as well as errors, warnings, or simply informative messages. It offers a great functionality by means of message (or logging) levels, conditional messages, interfaces for STL streams, and much more. To put things in a simple and straightforward fashion, in order to print an informative message (or information), we can do the following at any point in the code:

```
ROS_INFO( "My INFO message." );
```

Note that we do not have to include any particular library in our source code as long as the main ROS header is included. However, we can add the `ros/console.h` header as shown in the following code snippet:

```
#include <ros/ros.h>
#include <ros/console.h>
```

As a result of running a program with the preceding message, we will have the following output:

```
[ INFO] [1356440230.837067170]: My INFO message.
```

All messages are printed with its level and the current timestamp (your output might differ for this reason) before the actual message and both these values are between square brackets. The timestamp is the epoch time, that is, the seconds and nanoseconds since 1970 followed by our message.

This function allows parameters in the same way as the `printf` function in C. This means that we can pass values using all special characters that we can use with `printf`; for example, we can print the value of a floating point number in the variable `val` with this code:

```
const double val = 3.14;
ROS_INFO( "My INFO message with argument: %f", val );
```

Also, C++ STL streams are supported with `*_STREAM` functions. Therefore, the previous instruction is equivalent to the following using streams:

```
ROS_INFO_STREAM(
  "My INFO stream message with argument: " << val
);
```

Please note that we did not specify any stream because it is implicit that we refer to `cout` or `cerr`, depending on the message level, as we will see in the next section.

Setting the debug message level

ROS comes with five classic logging levels, which are in the order of relevance. They are `DEBUG`, `INFO`, `WARN`, `ERROR`, and `FATAL`.

These names are part of the function or macro used to output messages that follows this syntax:

ROS_<LEVEL>[_<OTHER>]

Both `DEBUG` and `INFO` messages go to `cout` (or `stdout`). Meanwhile, `WARN`, `ERROR`, and `FATAL` go to `cerr` (or `stderr`). Also, each message is printed with a particular color as long as the terminal has this capability. The colors are `DEBUG` in green, `INFO` in white, `WARN` in yellow, `ERROR` in red, and `FATAL` in purple.

The names of these messages clearly say the typology of the message given. The user must use them accordingly. As we will see in the following sections, this allows us to output only messages starting at a user-defined minimum level so that debugging messages can be omitted when our code is stable. Additionally, we have `OTHER` variants that are explained in the sequel.

> **Downloading the example code**
>
> You can download the example code files for all Packt books you have purchased from your account at `http://www.packtpub.com`. If you purchased this book elsewhere, you can visit `http://www.packtpub.com/support` and register to have the files e-mailed directly to you. You can also download these code files from `https://github.com/AaronMR/Learning_ROS_for_Robotics_Programming`.

Configuring the debugging level of a particular node

By default, only messages of INFO or a higher level are shown. ROS uses the levels to filter the messages printed by a particular node. There are many ways to do so. Some of them are set at the compile time, while others can be changed before execution using a configuration file. It is also possible to change this level dynamically, as we will see later in the following sections, using rosconsole and rxconsole.

First, we will see how to set the debugging level at compile time in our source code. Just go to the main function, and after the ros::init call, insert the following code:

```
// Set the logging level manually to DEBUG
ROSCONSOLE_AUTOINIT;
log4cxx::LoggerPtr my_logger =
  log4cxx::Logger::getLogger( ROSCONSOLE_DEFAULT_NAME );
my_logger->setLevel(
  ros::console::g_level_lookup[ros::console::levels::Debug]
);
```

You do not need to include any particular header, but in the CMakeLists.txt file, we must link the header to the log4cxx library. To do so, we must put:

```
find_library(LOG4CXX_LIBRARY log4cxx)
```

And our node must link to it:

```
rosbuild_add_executable(example1 src/example1.cpp)
target_link_libraries(example1 ${LOG4CXX_LIBRARY})
```

Now, DEBUG (and higher) messages are shown when our node runs since we set ros::console::levels::Debug in the preceding example. You can run the example1 node to check it and even change the level.

An alternative to the preceding method consists of using the compile-time-logger-removal macros. Note that this will remove all the messages below a given level at compilation time, so later we will not have them; this is typically useful for the release build of our programs. To do so, we must set ROSCONSOLE_MIN_SEVERITY to the minimum severity level or even none, in order to avoid any message (even FATAL); this macro can be set in the source code or even in the CMakeLists.txt file. The macros are ROSCONSOLE_SEVERITY_DEBUG, ROSCONSOLE_SEVERITY_INFO, ROSCONSOLE_SEVERITY_WARN, ROSCONSOLE_SEVERITY_ERROR, ROSCONSOLE_SEVERITY_FATAL, and ROSCONSOLE_SEVERITY_NONE.

The ROSCONSOLE_MIN_SEVERITY macro is defined in ros/console.h as DEBUG if not given. Therefore, we can pass it as a built argument (with -D) or put it before all the headers; for example, to show only ERROR (or higher) messages we will execute the following code as we did in the example2 node:

```
#define ROSCONSOLE_MIN_SEVERITY ROSCONSOLE_SEVERITY_DEBUG
```

On the other hand, we have a more flexible solution of setting the minimum debugging level in a configuration file. We will create a folder, just for convenience, named config with the file chapter3_tutorials.config and this content:

```
log4j.logger.ros.chapter3_tutorials=DEBUG
```

We can put any of the levels supported in ROS. Then, we must set the ROSCONSOLE_CONFIG_FILE environment variable to point our file. However, there is a better option. It consists of using a launch file that does this and also runs our node directly. Therefore, we can extend the launch files shown before to do so with an env element as shown in the following code snippet:

```
<launch>
  <!-- Logger config -->
  <env name="ROSCONSOLE_CONFIG_FILE"
       value="$(find chapter3_tutorials)/config/chapter3_tutorials.
config"/>

  <!-- Example 1 -->
  <node pkg="chapter3_tutorials" type="example1" name="example1"
        output="screen"/>
</launch>
```

The environment variable takes the config file, described previously, that contains the logging level specification for each named logger. Then, in the launch file, our node is simply run.

Giving names to messages

Since we can put messages in many places inside the same node, ROS allows us to give a name to each node in our program. This way, later on, it will be easier to detect from which part of the code is such a message coming. To do so, we use the ROS_<LEVEL>[_STREAM]_NAMED function as shown in the following code snippet (taken from the example2 node):

```
ROS_INFO_STREAM_NAMED(
  "named_msg",
  "My named INFO stream message; val = " << val
);
```

With named messages, we can go back to the `config` file and set different debugging levels for each named message. This allows for fine tuning using the name of the messages as the children of the node in the specification; for example, we can set the `named_msg` messages that are shown only for the `ERROR` (or higher) level with (note that although ROS uses log4cxx, the configuration files use the log4j root name) the following command line:

```
log4j.logger.ros.chapter3_tutorials.named_msg=ERROR
```

Conditional and filtered messages

Conditional messages are printed only when a given condition is satisfied. In some way, they are like conditional breakpoints using debugging messages. To use them, we have the `ROS_<LEVEL>[_STREAM]_COND[_NAMED]` functions; note that they can be named messages as well. The following are the examples of the `example2` node:

```
// Conditional messages:
ROS_INFO_STREAM_COND(
  val < 0.,
  "My conditional INFO stream message; val (" << val << ") < 0"
);
ROS_INFO_STREAM_COND(
  val >= 0.,
  "My conditional INFO stream message; val (" << val << ") >= 0"
);

// Conditional Named messages:
ROS_INFO_STREAM_COND_NAMED(
  "cond_named_msg", val < 0.,
  "My conditional INFO stream message; val (" << val << ") < 0"
);
ROS_INFO_STREAM_COND(
  "cond_named_msg", val >= 0.,
  "My conditional INFO stream message; val (" << val << ") >= 0"
);
```

Filtered messages are similar to conditional messages in essence, but they allow us to specify a user-defined filter that extends `ros::console::FilterBase`. We must pass a pointer to such a filter in the first argument of a macro with the format `ROS_<LEVEL>[_STREAM]_FILTER[_NAMED]`. The following example is taken from the `example2` node:

```
struct MyLowerFilter : public ros::console::FilterBase {
  MyLowerFilter( const double& val ) : value( val ) {}
```

```
  inline virtual bool isEnabled()
  {
    return value < 0.;
  }

  double value;
};

struct MyGreaterEqualFilter : public ros::console::FilterBase {
  MyGreaterEqualFilter( const double& val ) : value( val ) {}

  inline virtual bool isEnabled()
  {
    return value >= 0.;
  }

  double value;
};

MyLowerFilter filter_lower( val );
MyGreaterEqualFilter filter_greater_equal( val );

ROS_INFO_STREAM_FILTER(
  &filter_lower,
  "My filter INFO stream message; val (" << val << ") < 0"
);
ROS_INFO_STREAM_FILTER(
  &filter_greater_equal,
  "My filter INFO stream message; val (" << val << ") >= 0"
);
```

More messages – once, throttle, and combinations

It is also possible to control how many times a given message is shown. We can print it only once with `ROS_<LEVEL>[_STREAM]_ONCE[_NAMED]`. This kind of message is useful in loops where we do not want to output so many messages for efficiency reasons, and it is enough to know that we entered.

```
for( int i = 0; i < 10; ++i ) {
  ROS_INFO_STREAM_ONCE(
    "My once INFO stream message; i = " << i
  );
}
```

This code from the `example2` node will show the message only once for `i == 0`.

However, it is usually better to show the message at every iteration. This is where we can use throttle messages. They have the same format as that of the `ONCE` message, but if you replace `ONCE` with `THROTTLE` they will have `Duration` as the first argument; that is, it is printed only at the specified time interval:

```
for( int i = 0; i < 10; ++i ) {
  ROS_INFO_STREAM_ONCE(
    2,
    "My once INFO stream message; i = " << i
  );
  ros::Duration( 1 ).sleep();
}
```

Finally, note that named, conditional, and once/throttle messages can be combined together for all the available levels.

Nodelets also have some support in terms of logging and debugging messages. Since they have their own namespace, they have a specific name to differentiate the messages of one nodelet from another. Simply, all the macros shown so far are valid, but instead of `ROS_*` we have `NODELET_*`. These macros will only compile inside nodelets. Also, they operate by setting up a named logger, with the name of the nodelet running, so that you can differentiate between the output of two nodelets of the same type running in the same nodelet manager. Another advantage of nodelets is that they will help you turn one specific nodelet to the `DEBUG` level, instead of all the nodelets of a specific type.

Using rosconsole and rxconsole to modify the debugging level on the fly

A logging message integrates with a series of tools to visualize and configure them. The ROS framework comes with an API known as **rosconsole** that was used to some extent in the previous sections. We advised you to consult the API for more advanced features, but we believe that this book covers everything a regular robotics user or developer might need.

Apart from the API to configure the logging message in your nodes, ROS provides a graphical tool, which is part of the `rxtools` package. This tool is `rxconsole`, and you only have to type it in the command line to see the graphical interface that allows seeing, inspecting, and configuring the logging subsystem of all running nodes.

In order to test this, we are going to use `example3`, so we will run `roscore` in one terminal and the node in another using the following command line:

```
rosrun chapter3_tutorials example3
```

Now, with the node running, we will open another terminal and run rxconsole. The following window will open; note that you can also run rxconsole first. So the logging message will now appear immediately as shown in the following screenshot:

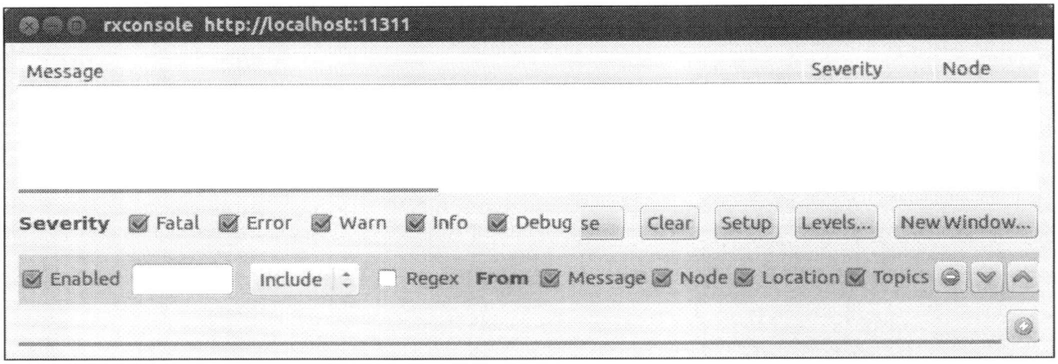

Once we have our `example3` node running, we will start to see messages as shown in the following screenshot:

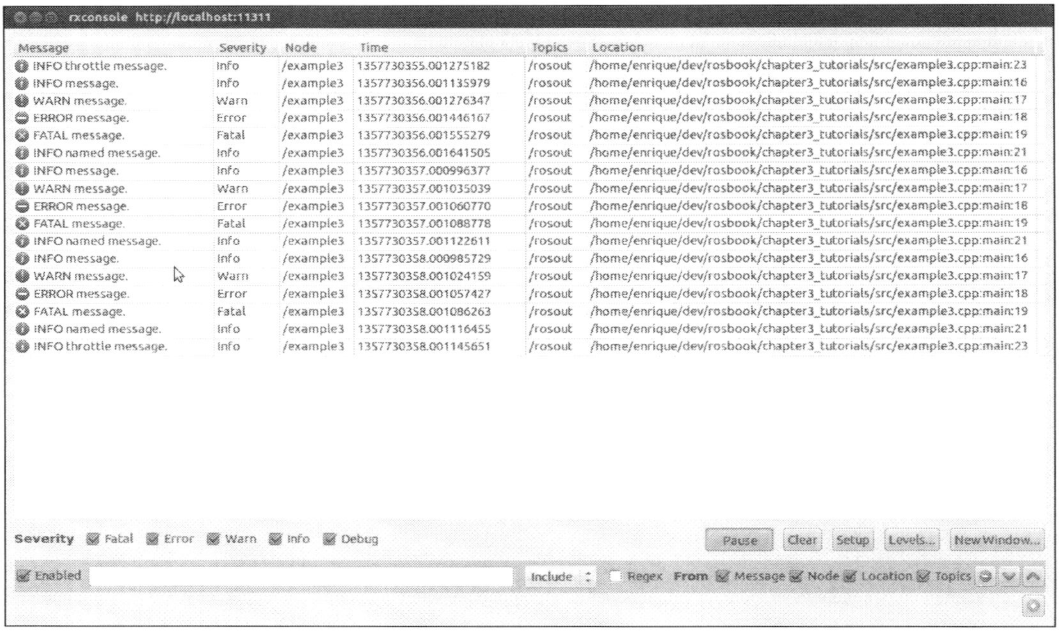

In the table, we have several columns that give information (aligned in order, as shown in the preceding screenshot) about the message itself, its severity, the node that generated the message and the timestamp, the topic (the `/rosout` aggregator from the ROS server is usually with it), and the location of the logging macro in the source code of the node.

We can click on the **Pause** button to stop receiving new messages, and click on it again to resume monitoring. For each message in the table, we can click on **Pause** to see all its details as shown in the following screenshot for the last message of the previous screenshot:

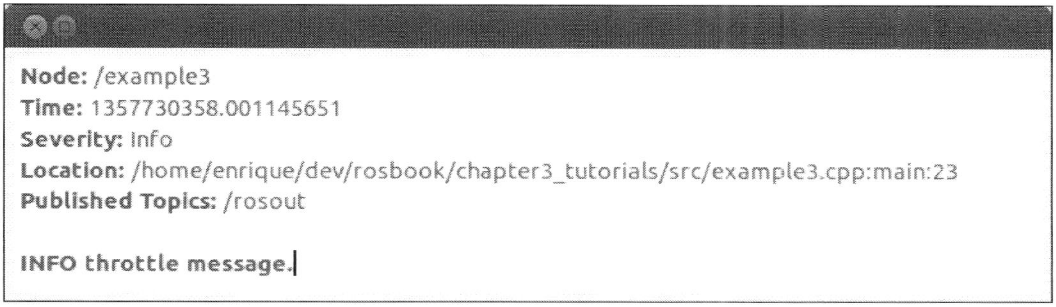

One of the great features of rxconsole is the ability to filter messages. The most basic way to filter is by severity level. If we want to see only the `FATAL` and `ERROR` messages in our example, we have to unselect the other severity levels as shown in the following screenshot:

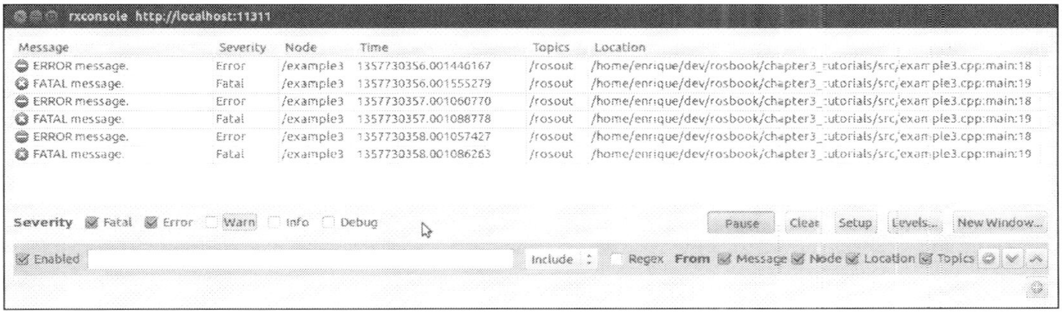

Immediately, we will see only the messages with `FATAL` and `ERROR` severity levels. Additionally, we can filter (include or exclude) by message content, node name, location, or topic name, and also using regular expression in our queries. The following screenshot shows an example in which we filtered the messages to show only those with the word `named` in the message for all the severity levels:

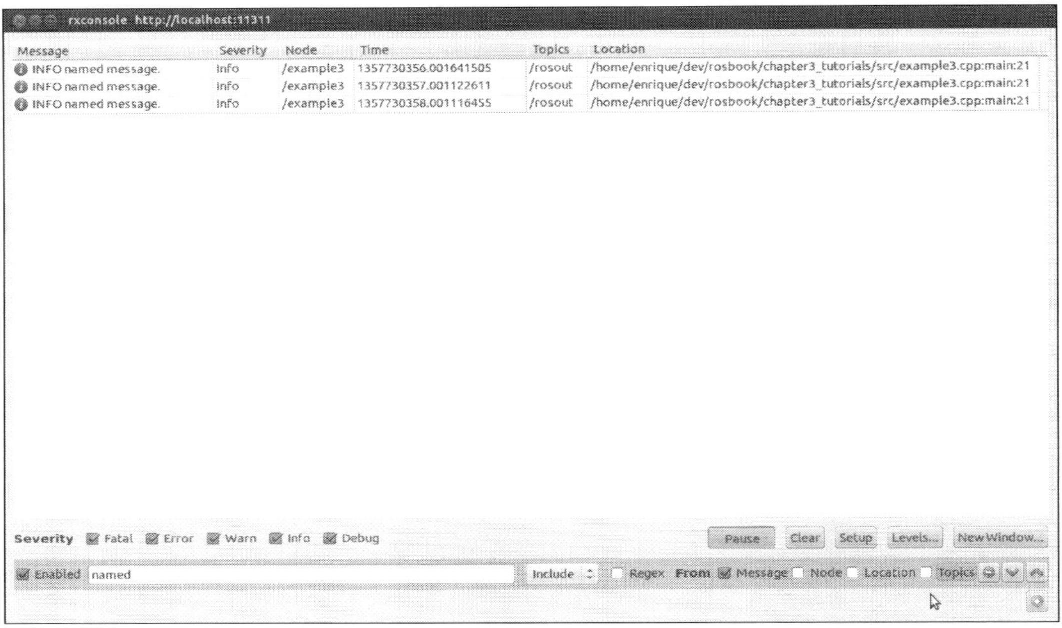

Note that we can add more filter entries (as much as we want) and also remove or disable the ones defined.

We can also remove all the messages captured by rxconsole by clicking on the **Clear** button. The **Setup** button allows configuring of the diagnostic aggregator topic, which is typically `rosout_agg`, and the number of messages that the GUI keeps in its history before rolling over (as shown in the following screenshot). Note that the diagnostic aggregator is just a sink in the ROS server that receives all the logging messages. This way, we can inspect which devices are failing or how are they working. An advanced developer might find it useful to learn about the diagnostic API to use a higher layer to build hierarchical layout that is already supported and used for complex systems or robots.

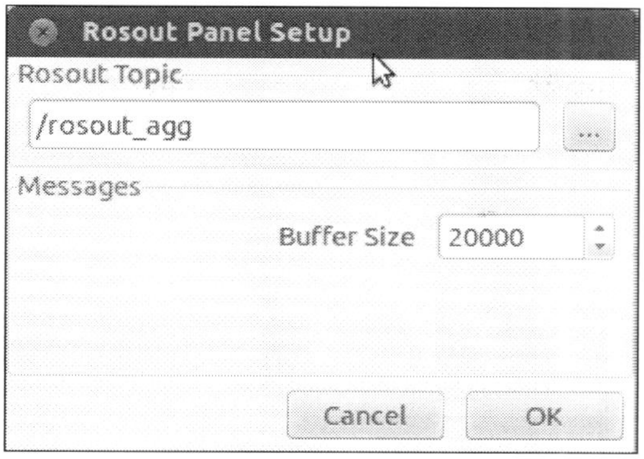

Finally, we can set the logger severity level for each named logger. By default, each node has a logger with its package name, but we also can define named loggers using the `NAMED` macros described in the previous sections. For the `example3` node, we have the `ros.chapter3_tutorials` and `ros.chapter3_tutorials.named_msg` loggers. If we click on the **Levels...** button, we will see the following screenshot:

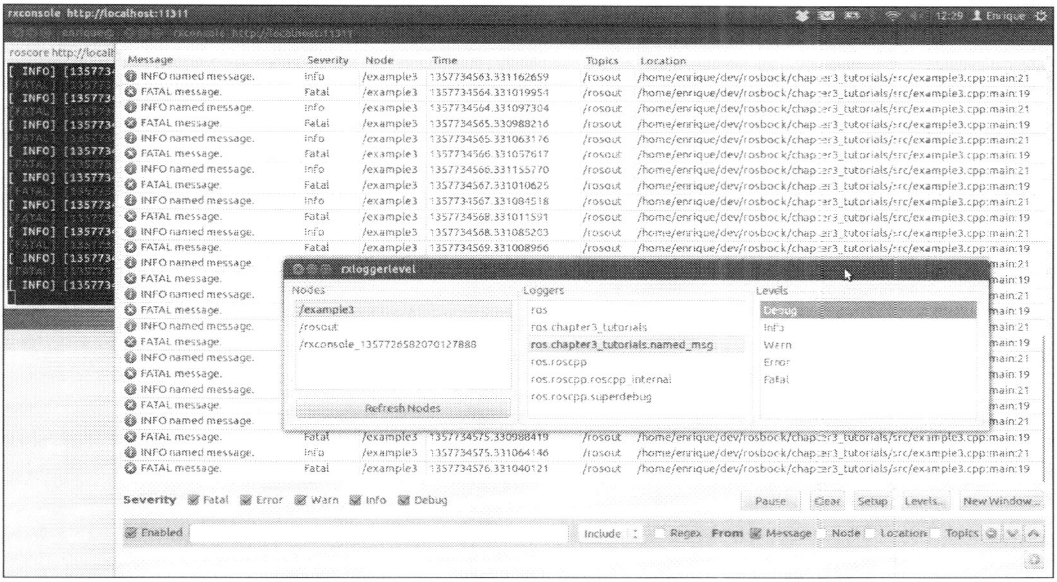

Here, we can select the `example3` node and then the logger in order to set its level. Note that there are some internal loggers that you can use apart from the ones mentioned before. In the preceding screenshot, we set the `DEBUG` severity level for the `ros.chapter3_tutorials.named_msg` logger so that all the messages with this level or higher are shown; that is, all messages in this case.

Inspecting what is going on

When our system is running, we might have several nodes and even more topics publishing messages and connected by subscription among nodes. Also, we might have some nodes providing services as well. For large systems, it is important to have some tools that let us see what is running at a given time. ROS provides us with some basic but powerful tools to do so, and also to detect a failure in any part of the nodes graph; that is, the architecture that emerges from the connection of ROS nodes using topics.

Listing nodes, topics, and services

In our honest opinion, we should start with the most basic level of introspection. We are going to see how to obtain the list of nodes running, and topics and services available at a given time. Although extremely simple, this is very useful and robust.

- To obtain the list of nodes running use:
  ```
  rosnode list
  ```
- The topics of all nodes are listed using:
  ```
  rostopic list
  ```
- And, similarly, all services are shown by:
  ```
  rosservice list
  ```

We recommend you to go back to *Chapter 2, The ROS Architecture with Examples* to see how these commands also allow you to obtain the message type sent by a particular topic, as well as its fields, using `rosmsg show`.

Inspecting the node's graph online with rxgraph

The simplest way to illustrate the current state of an ROS session is with a directed graph that shows the nodes running on the system and the publisher-subscriber connections among these nodes through the topics. The ROS framework provides some tools to generate such a node's graph. The main tool is rxgraph, which shows the node's graph during the system execution and allows us to see how a node appears or disappears dynamically.

To illustrate how to inspect the nodes, topics, and services with rxgraph, we are going to run the `example4` and `example5` nodes simultaneously with the following file:

```
roslaunch chapter3_tutorials example4_5.launch
```

The `example4` node publishes in two different topics and calls a service. Meanwhile, `example5` subscribes to those topics and also has the service server attend the request queries and provide the response. Once the nodes are running, we have the node's topology as shown in the following screenshot:

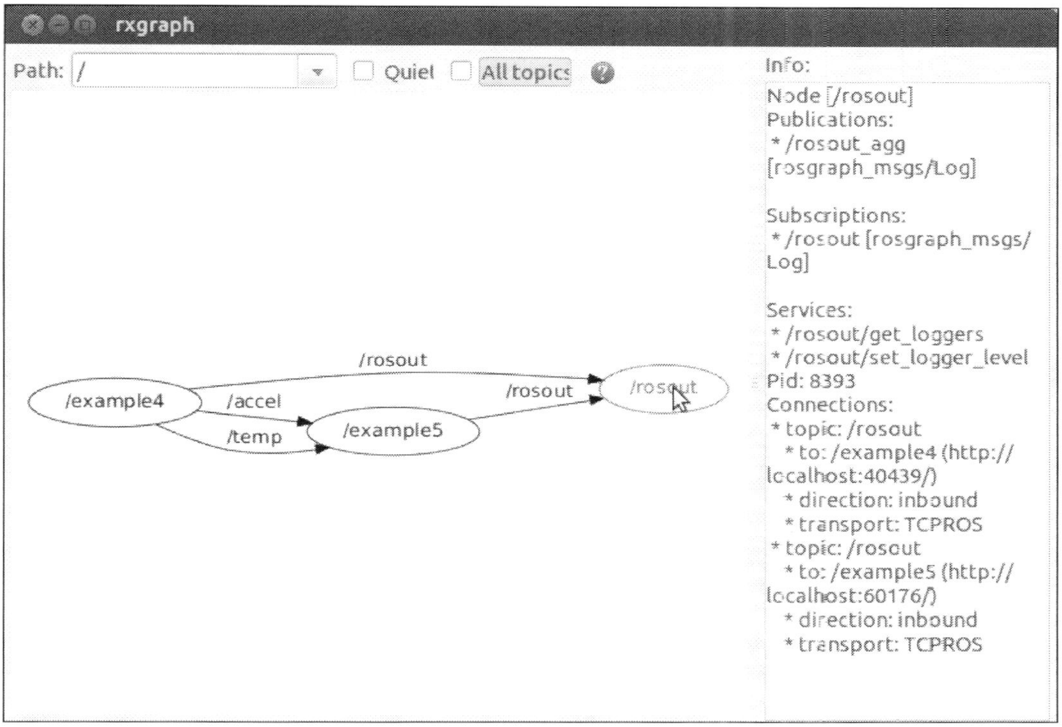

In the preceding screenshot, we have nodes connected by topics. We also have the ROS server node as **rosout**, as well as the rosout topics that publish the log message for the diagnostic aggregator in the server as we have seen previously. There is also a panel to the right with information regarding the node selected. In the preceding screenshot, we have information regarding the server; that is, its IP and port for the remote nodes, topics, and connections.

We can enable the quiet view so that the ROS server is omitted. This is useful for large systems because it is always there but does not provide any information on our system's topology itself. To see the node service, we turn on the node and will see the right-hand panel. In the following screenshot, we have highlighted the `speed` service of the `example5` node:

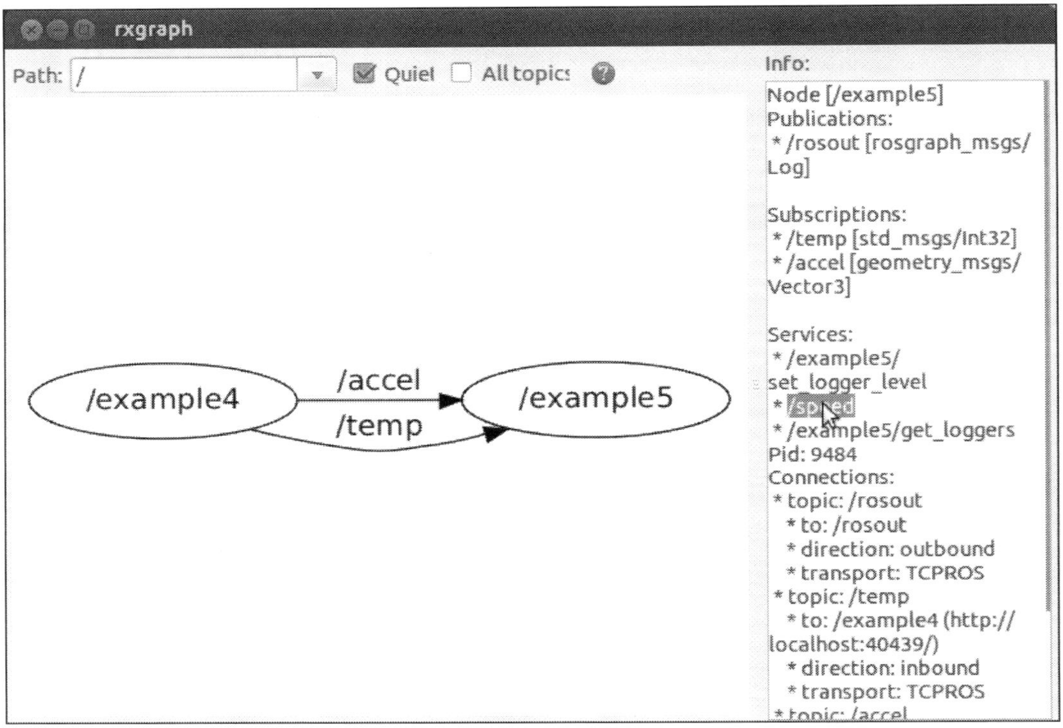

When there is a problem in the system, the nodes appear in red all the time (not just when we hover the mouse over). In such cases, it is useful to select **All topics** to show unconnected topics as well. Sometimes, the problems are a consequence of a misspelled topic name.

When running nodes in different machines, rxgraph shows its great high-level debugging capabilities since it shows whether the nodes see each other from one machine to the other.

When something weird happens – roswtf!

ROS also has another tool to detect potential problems in all the elements of a given package. Just move with `roscd` to the package you want to analyze, and then run `roswtf`. In the case of `chapter3_tutorials`, we have the following output:

```
user@pc$ roswtf
Loaded plugin tf.tfwtf
Package: chapter3_tutorials
================================================================================
Static checks summary:

Found 1 warning(s).
Warnings are things that may be just fine, but are sometimes at fault

WARNING ROS_IP may be incorrect: ROS_IP [localhost] does not appear to be a local IP address ['127.0.0.1', '192.168.1.35'].

Found 1 error(s).

ERROR The following packages have rpath issues in manifest.xml:
 * chapter3_tutorials: found flag "-L/opt/ros/fuerte/lib", but no matching "-Wl,-rpath,/opt/ros/fuerte/lib"
 * geometry_msgs: found flag "-L/opt/ros/fuerte/lib", but no matching "-Wl,-rpath,/opt/ros/fuerte/lib"
 * std_msgs: found flag "-L/opt/ros/fuerte/lib", but no matching "-Wl,-rpath,/opt/ros/fuerte/lib"
 * roscpp: found flag "-L/opt/ros/fuerte/lib", but no matching "-Wl,-rpath,/opt/ros/fuerte/lib"

================================================================================
Beginning tests of your ROS graph. These may take awhile...
analyzing graph...
... done analyzing graph
running graph rules...
... done running graph rules

Online checks summary:

Found 3 error(s).

ERROR Communication with [/example4] raised an error:
ERROR Communication with [/rosout] raised an error:
ERROR The following nodes should be connected but aren't:
 * /example4->/rosout (/rosout)
```

Normally, we should expect no error or warning but some of them are innocuous. In the preceding screenshot, we see that `roswtf` has detected that it was not able to connect with the `example4` node. This happens because this node has a `sleep` instruction, and if analyzed, this might occur while sleeping. The other errors are a consequence of this one. The purpose of `roswtf` is to signal potential problems, and then we are responsible for checking whether they are real or meaningless ones, as in the previous case.

Plotting scalar data

Scalar data can easily be plotted with some generic tools already available in ROS. Scalar data cannot be plotted, rather each scalar field has to be plotted separately. This is the reason we talk about scalar data because most nonscalar structures are better represented with ad hoc visualizers, some of which we will see later; for instance, images, poses, and orientation/attitude.

Creating a time series plot with rxplot

In ROS, scalar data can be plotted as a time series over the time provided by the timestamps of the messages. Then, we will plot our scalar data in the y axis. The tool to do so is `rxplot`. It has a powerful argument syntax that allows us to specify several fields of a structured message (in a concise manner as well).

To show `rxplot` in action, we are going to use the `example4` node since it publishes a scalar and a vector (nonscalar) in two different topics, which are `temp` and `accel` respectively. The values put in these messages are synthetically generated, so they have no actual meaning but are useful for plotting demonstration purposes. So, start by running the node with:

```
rosrun chapter3_tutorials example4
```

With `rostopic list`, you will see the topics `temp` and `accel` available. Now, instead of the typical `rostopic echo <topic>`, we will use `rxplot` so that we can see the values graphically over time.

To plot a message, we must know its format; use `rosmg show <msg type>` if you do not know it. In the case of scalar data, we always have a field called `data` that has the actual value. Hence, for the `temp` topic, which is of the type `Int32`, we will use:

```
rxplot /temp/data
```

With the node running, we will see a plot that changes over time with incoming messages as shown in the following screenshot:

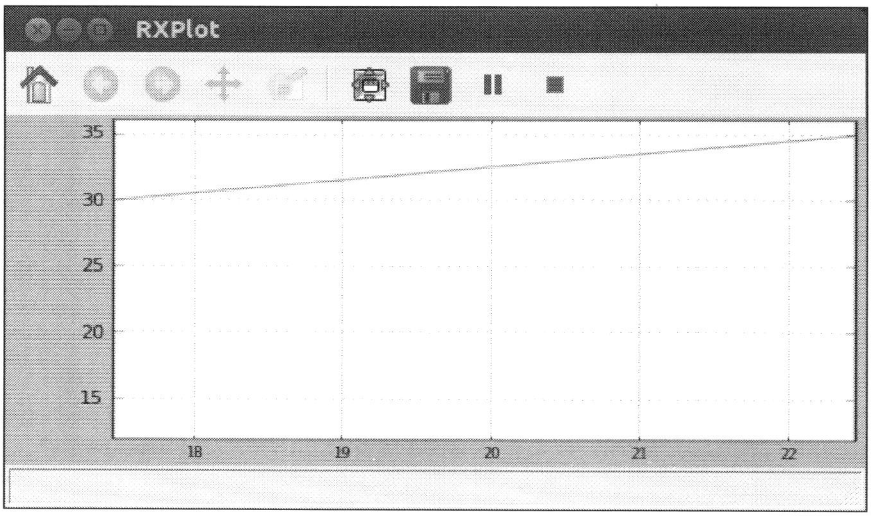

For the `accel` topic provided by the example node, in which we have a `Vector3` message (as you can check with `rostopic type /accel`), we can plot three fields of the vector in a single plot, which is a great feature of `rxplot`. The `Vector3` message has the fields x, y, and z. We can specify the fields separated by commas (,) or in a more concise manner as follows:

```
rxplot /accel/x:y:z
```

The plot will look like this:

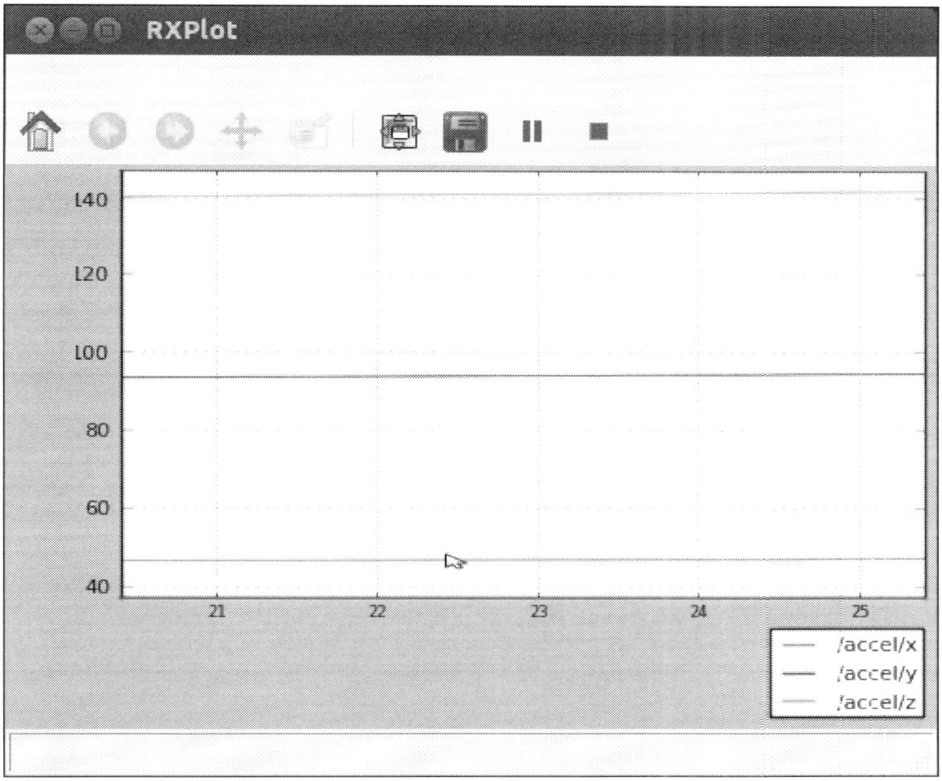

We can also plot each field in separate axes as shown in the next screenshot. To do so, we separate each field by a blank space; remember that when you use commas, you must not insert any spaces. Therefore, if we run `rxplot /accel/x /accel/y /accel/z`, the plot will show like this:

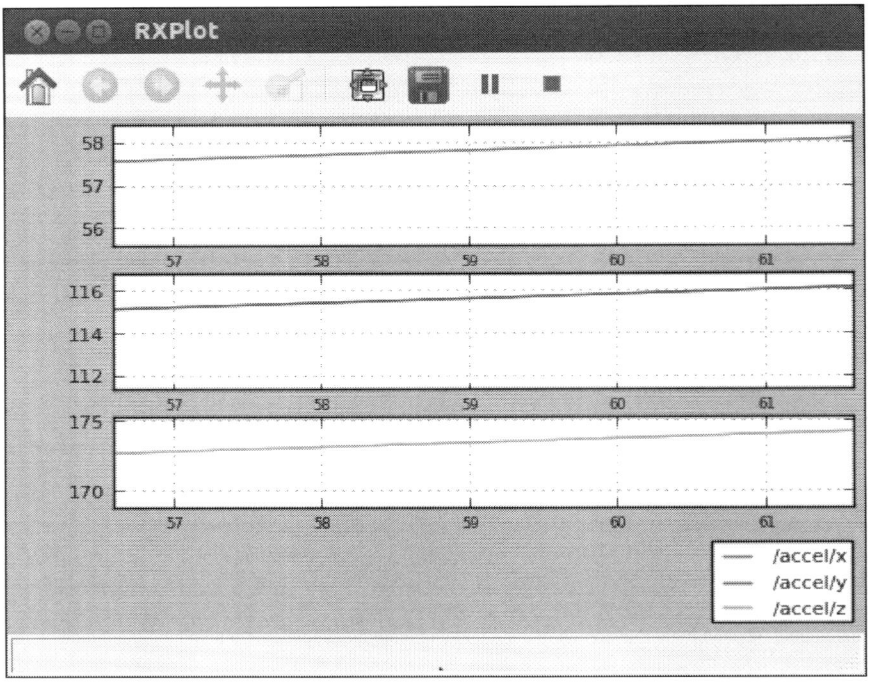

Other plotting utilities – rxtools

The `rxplot` tool is part of the `rxtools` package along with other tools. You might go to this package to see more GUI or batch tools that can help in the development of robotic applications and the process of debugging, monitoring, and introspecting. It is also important to know that being a node (inside this package), these tools can also be run from a `launch` file.

In the case of `rxplot`, in order to run it from a `launch` file, we must put the following code inside it:

```
<node pkg="rxtools" type="rxplot" name="accel_plot"
        args="/accel/x:y:z"/>
```

Note that we use the `args` argument of the `node` element in the `launch` file to pass `rxplot` the arguments.

Visualization of images

In ROS, we have a node that allows us to show images coming from a camera on the fly. You only need a camera to do this. It is also possible to reproduce a video with a simple node in ROS but here we are going to use your laptop's webcam. The `example6` node implements a basic camera capture program using OpenCV and ROS bindings to convert `cv::Mat` images into ROS image messages that can be published in a topic. This node publishes the camera frames in the `/camera` topic.

We are only going to run the node with a launch file created to do so. The code inside the node is still new for the reader, but in the upcoming chapters, we will cover how to work with cameras and images in ROS so that we can come back to this node and understand every bit of the code:

```
roslaunch chapter3_tutorials example6.launch
```

Once the node is running, we can list the topics (`rostopic list`) and see that the `/camera` topic is there. A straightforward way to see that we are actually capturing images is to see at which frequency we are receiving images in the topic with `rostopic hz /camera`. It should be something like 30 Hz usually, but at least some value must be seen:

```
subscribed to [/camera]
average rate: 30.131
min: 0.025s max: 0.045s std dev: 0.00529s window: 30
```

Visualizing a single image

Being an image, we cannot use `rostopic echo /camera` because the amount of information in plain text would be very huge and also difficult to analyze. Hence, we are going to use the following code:

```
rosrun image_view image_view image:=/camera
```

This is the `image_view` node, which shows the images in the given topic (the `image` argument) in a window, as shown in the following screenshot:

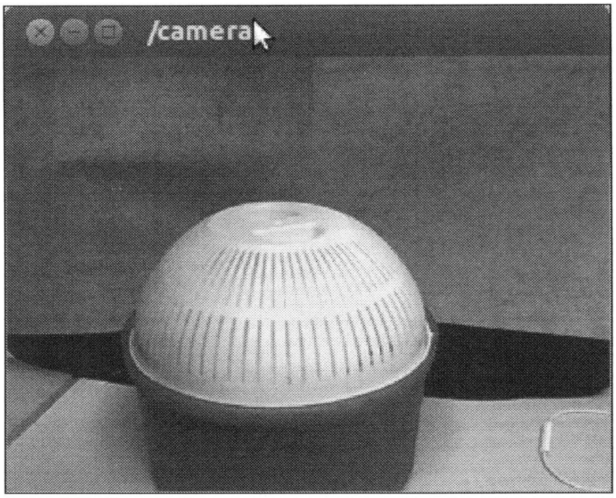

This way we can visualize every image or frame published in a topic in a very simple and flexible manner, even over a network. If you right-click on the window, you can save the current frame in the disk, usually in your home directory or `~/.ros`.

FireWire cameras

In the case of FireWire cameras, ROS also provides a larger set of tools that support calibration, both mono and stereo, as well as a way to change the camera parameters dynamically with the `reconfigure_gui` node. Usually, FireWire cameras allow changing some configuration parameters of the sensor, such as the frame rate, shutter speed, and brightness. ROS already comes with a driver for FireWire (IEEE 1394, a and b) cameras that can be run using the following command:

```
rosrun camera1394 camera1394_node
```

Once the camera is running, we can configure its parameters with the `reconfigure_gui` node, in which the first thing we do is the selection of the node we want to configure. We only have to run this:

```
rosrun dynamic_reconfigure reconfigure_gui
```

We will see an interface with all the configuration parameters and a series of slider or comboboxes, depending on the data type, to set its value within the valid limits. The following screenshot illustrates this for a particular FireWire camera:

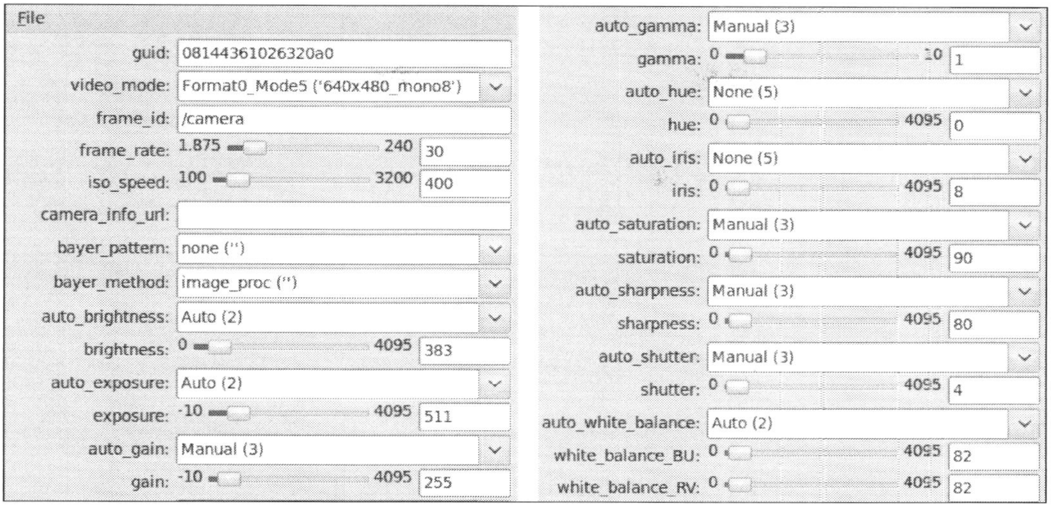

Note that we will cover how to work with cameras in later chapters. Also, note that the parameters reconfiguration, from the developer point of view, will be explained in detail in *Chapter 6, Computer Vision*.

Another important utility that ROS gives to the user is the possibility to calibrate the camera. It has a calibration interface built on top of the OpenCV calibration API. We will also cover this in *Chapter 6, Computer Vision*, when we see how to work with cameras. This tool is very simple to use; so, once the camera is running, we only have to show some views of a calibration pattern (usually a checkerboard) using this:

```
rosrun camera_calibration cameracalibrator.py --size 8x6 --square 0.108
image:=/camera/image_raw camera:=/camera
```

You can see the checkerboard pattern in the following screenshot:

After the calibration, the result will be the so-called camera matrix and distortion coefficients along with the views used to compute it. Although we will see this later in the book, it is important to say that in the case of the FireWire camera, all the calibration information is saved in a file that is pointed by the camera configuration. Hence, the ROS system allows seamless integration so that we can use the `image_proc` tool to rectify the images; that is, to correct the distortion as well as to de-Bayer the raw images if they were in Bayer.

Working with stereo vision

To some extent, ROS also supports stereo vision. If you have a stereo pair, you can calibrate both cameras simultaneously as well as the baseline between them. For this, we will use:

```
rosrun camera_calibration cameracalibrator.py --size 8x6 --square 0.108
right:=/my_stereo/right/image_raw left:=/my_stereo/left/image_raw right_
camera:=/my_stereo/right left_camera:=/my_stereo/left
```

This command runs the Python camera calibrator node and receives two initial parameters that indicate the type of calibration pattern. In particular, the size specified is the number of inner corners (8 x 6 in the example) and the dimensions of the square cells. Then, the topics that publish the right and left raw images and the respective camera information topics are given. In this case, the interface will show the images from the left and right cameras and the results will be for each as well. They will include the baseline as well that is useful for some stereo tools.

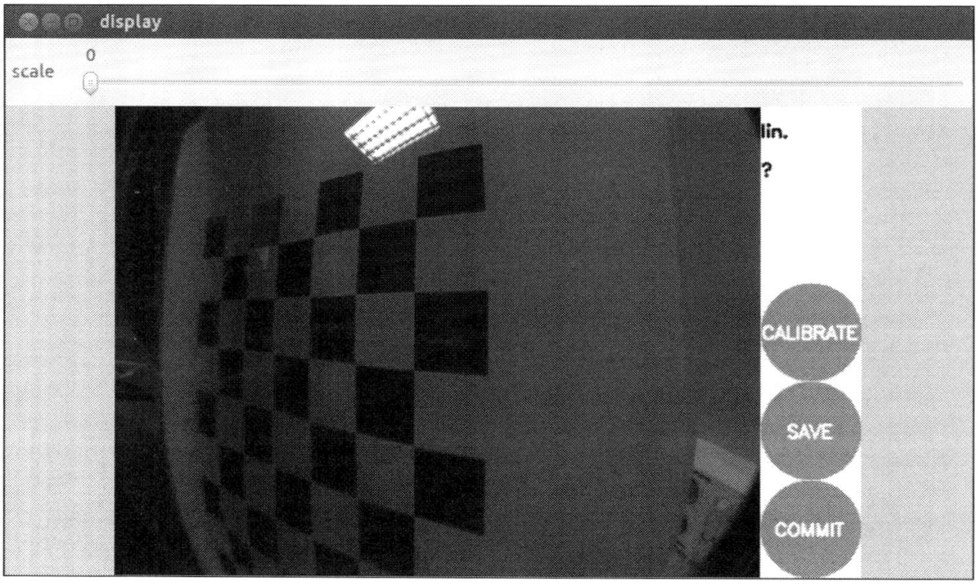

The stereo-specific tools allow you to compute the disparity image (refer to the following image), which is actually a way to obtain a 3D point cloud that represents the depth of each pixel in the real world. Therefore, the calibration of the camera and their baseline gives a 3D point cloud, up to some error and noise distribution that represents the real 3D position of each pixel in the world along with its color (or texture).

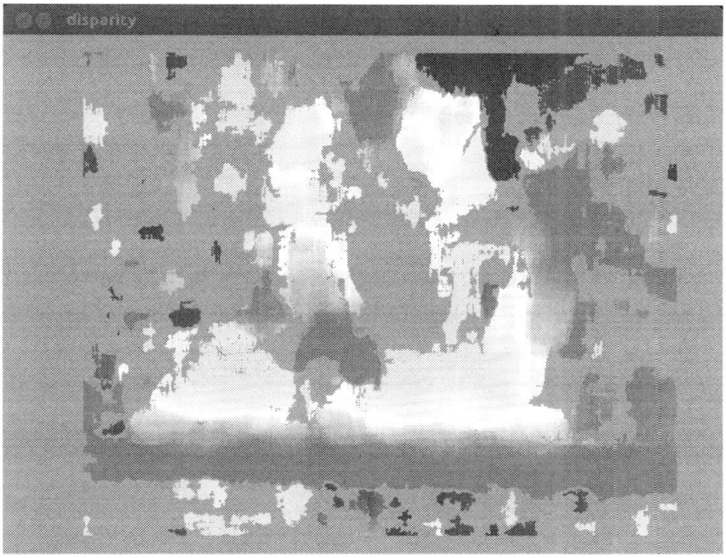

Similar to monocular cameras, we can generate the disparity image using stereo along with the rectified left and right images; in this case, using `stereo_image_proc`.

3D visualization

As we have seen in the previous section, there are some devices (such as stereo cameras, 3D laser, and the Kinect sensor) that provide 3D data, usually in the form of point clouds (organized or not). For this reason, it is extremely useful to have tools that visualize this type of data. In ROS, we have rviz, which we will see in the following section, that integrates an OpenGL interface with a 3D world that represents sensors' data in a modeled world. To do so, we will see that, for complex systems, the frame of each sensor and the transformations among them is of crucial importance.

Visualizing data on a 3D world using rviz

With `roscore` running, we only have to execute the following code to start rviz:

```
rosrun rviz rviz
```

We will see the graphical interface in the following screenshot:

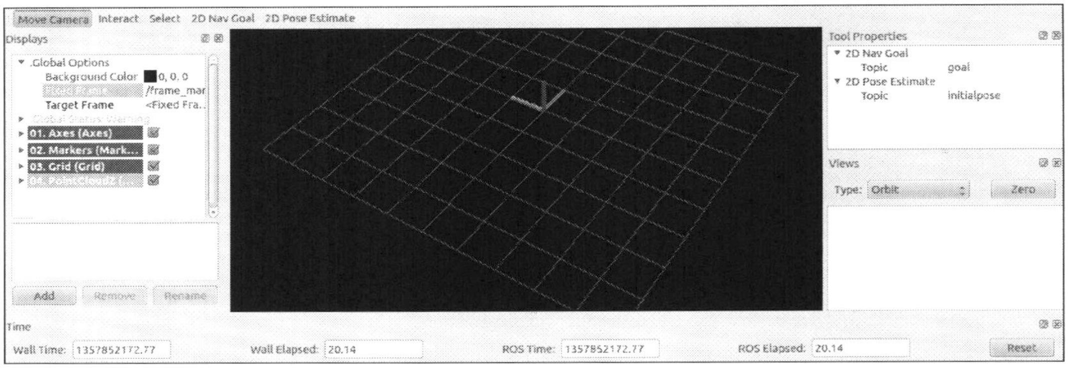

To the left, we have the **Displays** pane, in which we have a tree list of the different elements in the world, which appears in the middle. In this case, we have some elements that are already loaded. Indeed, this configuration or layout is saved in the `config/example7.cvg` file, which can be loaded by navigating to **File | Open Config**.

Below the **Displays** area, we have the **Add** button that allows the addition of more elements. Also, note that there are some global options, which are basically tools to set the fixed frame in the world with respect to which others might move. Then, we have **Axes (Axes)** and **Grid (Grid)** as a reference for the rest of the elements. In this case, for the `example7` node, we are going to see **Markers (Markers)** and **PointCloud2 (PointCloud2)**.

Finally, at the status bar, we have information regarding time, and to the right are the menus for the way to navigate in the world and select and manipulate elements.

Now we are going to run the `example7` node:

```
roslaunch chapter3_tutorials example7.launch
```

In rviz, we are going to set `frame_id` of the marker, that is `frame_marker`, in the fixed frame. We will see a red cube marker moving as shown in the following screenshot:

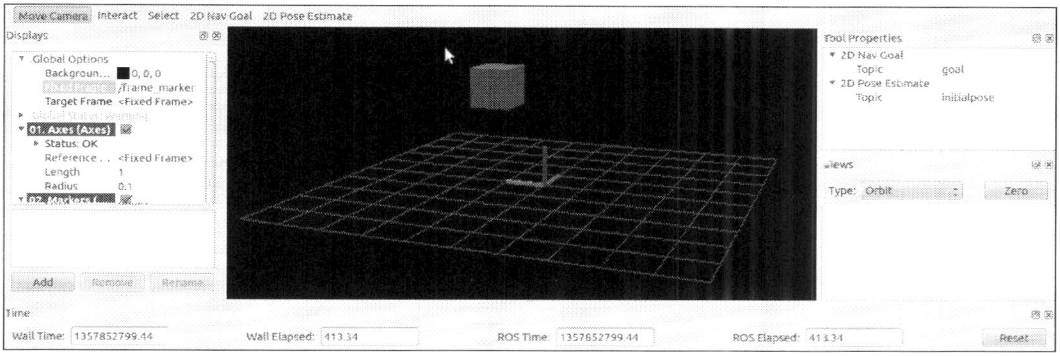

Similarly, if we set the fixed frame to `frame_pc`, we will see a point cloud that represents a plane of 200 x 100 points as shown in the following screenshot:

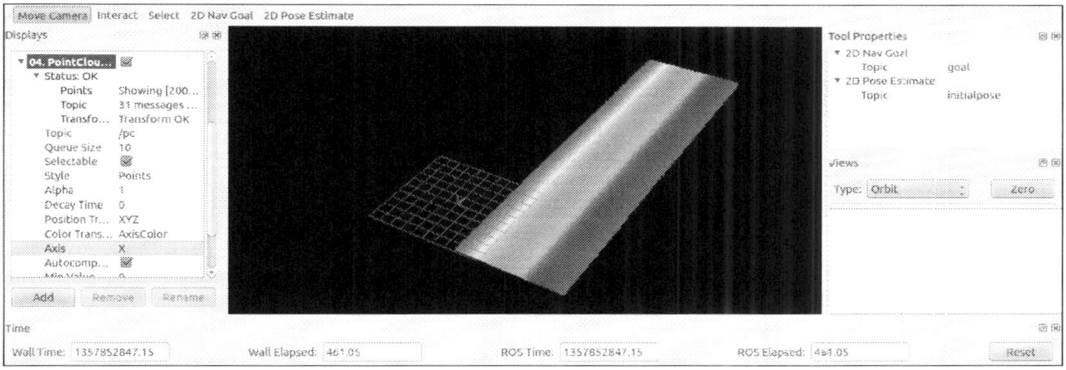

The list of supported built-in types in rviz also includes cameras and images, which are shown in a window similar to **image_view**. In the case of the camera, its calibration is used and in the case of stereo images it allows us to overlay the point cloud. We can also see laser scan data from range lasers and range cone values from IR/SONAR sensors.

Basic elements can also be represented, such as a polygon, several kinds of markers, a map (usually a 2D occupancy grid map), and even interactive marker objects, which allow users to set a pose (position and orientation) in the 3D world.

For the navigation stack that we will cover in the next chapters, we have several data types that are also supported such as odometry (plots the robot odometry poses), and path (draws the path followed by the robot), which is a succession of pose objects. Among other types, it is also worth mentioning the robot model, which shows the CAD model of all the robot parts, taking into account the transformation among the frame of each element. Indeed, **TF** (**transform frame**) elements can also be drawn, which is very useful for debugging the frames in the system; we will see an example in the next section.

This 3D graphical interface can also be embedded in the new `rqt_gui` GUI. We can also develop plugins for new types, and much more. However, we believe that the information given here is usually enough, and we recommend you consult the rviz documentation for further details and advanced topics.

The relationship between topics and frames

All topics must have a frame if they are publishing data from a particular sensor that have a physical location in the real world; for example, an accelerometer that is located in some position with respect to the mass center of the robot. If we integrate the accelerations to estimate the robot's velocities or its pose, we must take the transformation between the base (mass center) and the accelerometer frames. In ROS, the messages with a header, apart from the timestamp (also extremely important to put or synchronize different messages), can be assigned `frame_id`, which gives a name to the frame it belongs to.

But the frames itself are not very useful when we have more than a single device in our robot, each in a different frame/pose. We need the transformation among them. Actually, we have a frame transformation tree that usually has the base frame as its root. Then, we can see in rviz how this and other frames move with respect to the world frame.

Visualizing frame transformations

To illustrate how to visualize the frame transformations, we are going to use the turtlesim example. Run the following `launch` file then:

```
roslaunch turtle_tf turtle_tf_demo.launch
```

This is a very basic example with the purpose of illustrating the TF visualization in rviz. Note that for the different possibilities offered by the TF API, you should refer to later chapters of this book, in particular *Chapter 7, Navigation Stack – Robot Setups* and *Chapter 8, Navigation Stack – Beyond Setups*. For now, it is enough to know that they allow making the computations in one frame and then transforming them to another, including time delays.

It is also important to know that TFs are published at a certain frequency in the system, so it is like a subsystem where we can traverse the TF tree to obtain the transformation between any frames in it, and we can do it in any node of our system just by consulting TF.

If you receive an error, it is probably because the listener died on the launch startup, as another node that was required was not yet ready; so, please run the following on another terminal to start it again:

```
rosrun turtle_tf turtle_tf_listener
```

Now you should see a window with two turtles (the icon might differ) where one follows the other. You can control one of the turtles with the arrow keys but with the focus on the terminal for which the `launch` file is run. The following screenshot shows how one turtle has been following the other, after moving the one we can control for some time:

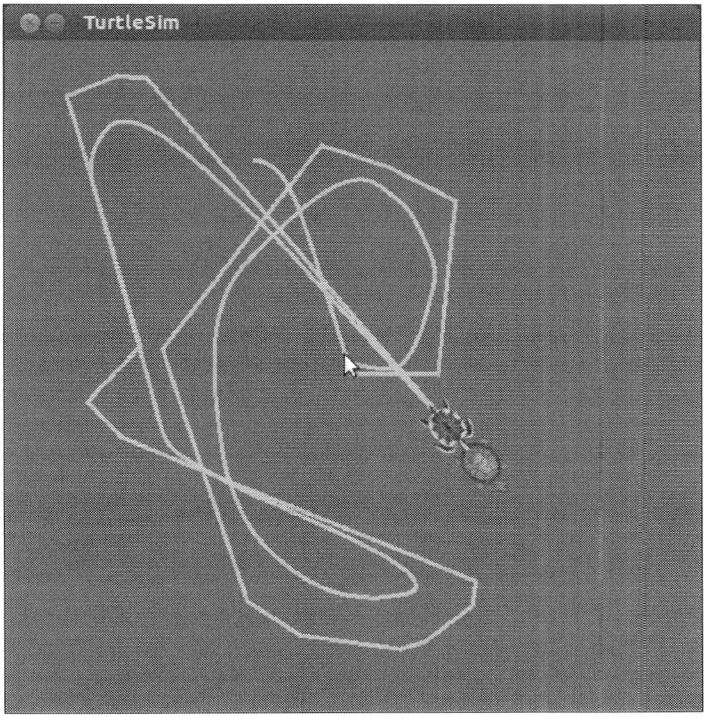

Each turtle has its own frame. We can see them in `rviz`:

```
rosrun rviz rviz
```

Now, instead of **TurtleSim**, we are going to see how the turtles' frames move in `rviz` while we move our turtle with the arrow keys. We have to set the fixed frame to `/world`, and then add the TF tree to the left area. We will see that we have the `/turtle1` and `/turtle2` frames, both as children of the `/world` frame. In the world representation, the frames are shown as axes. The `/world` frame is fixed because we configured it as such in `rviz`. It is also the root frame and the parent of the turtles' frames. This is represented with a yellow arrow that has a pink end. Also, set the view of the world to `TopDownOrtho` because this makes it easier to see how the frames move in this case, as they move only on the ground (2D plane). Also, you may find it useful to translate the world center, which is done with the mouse, as you do to rotate, but with the *Shift* key pressed.

In the following screenshot, you can see how the two frames of each turtle are shown with respect to the `/world` frame. We advise the user to play with the example to see it in action, in real time. Also, you might change the fixed frame. Note that `config/example_tf.vcg` is provided to give the basic `rviz` configuration used in this example.

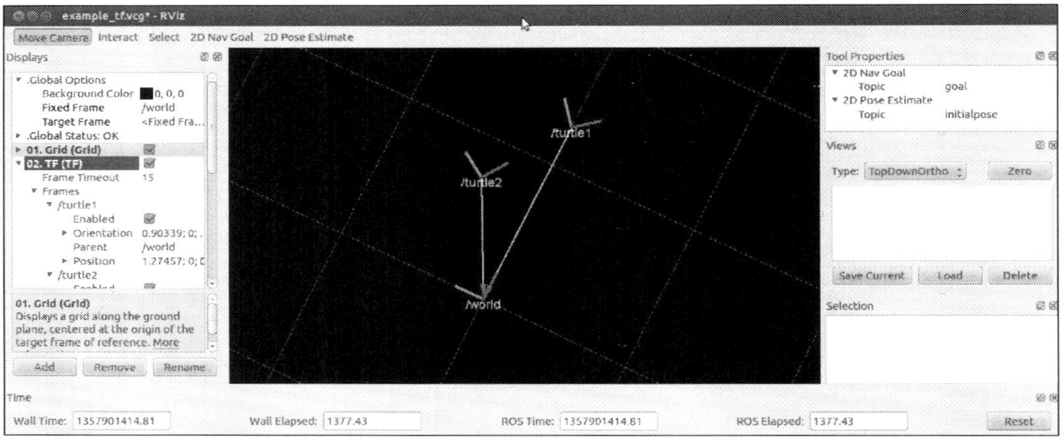

Saving and playing back data

Usually, when we work with robotic systems, the resources are shared, not always available, or the experiments cannot be done regularly because of the cost or time required to prepare and perform them. For this reason, it is good practice to record the data of the experiment session for future analysis and to work, develop, and test our algorithms. However, the process of saving good data so that we can reproduce the experiment offline is not trivial. Fortunately, we have powerful tools in ROS that already solve this problem.

ROS can save all messages published by the nodes through the topics. It has the ability to create a bag file that contains the messages as they are with all their fields and timestamps. This allows reproducing the experiment offline and simulating the real condition, which is the latency of message transmission. Moreover, ROS tools do all this efficiently with a high bandwidth and an adequate manner to organize the saved data.

In the next section, we explain the tools provided by ROS to save and playback the data stored in bag files, which use a binary format designed for and by ROS developers. We will also see how to manage these files; that is, inspect the content (number of messages, topics, and so on), compress, and split or merge several of them.

What is a bag file?

A bag file is a container of messages sent by topics that are recording during a session using a robot or some nodes. In brief, they are the logging files for the messages transferred during the execution of our system and allow us to playback everything even with the time delays, since all messages are recorded with a timestamp; not only for the timestamp in the header but also for the packets that have it. The difference between the timestamp used for recording and the one in the header is that the first one is set by the message that is recorded, while the other is set by the producer/publisher of the message.

The data stored in a bag file is in the binary format. The particular structure of this container allows for an extremely fast recording bandwidth, which is the most important concern while saving data. Also, the size of the bag file is relevant but is usually at the expense of speed. Anyway, we have the option to compress the file on the fly with the **bz2** algorithm; just use the `-j` parameter when you record with `rosbag record`, as you will see in the following section.

Every message is recorded along with the topic that published it. Therefore, we can specify which topics to record or just mention all (with `-a`). Later, when we play the bag file back, we can also select a particular subset of topics of all the ones in the bag file by indicating the names of the topics we want to be published.

Recording data in a bag file with rosbag

The first thing we have to do to start is simply record the data. We are going to use a very simple system, our `example4` node, as an example. Hence, we first run the node:

```
rosrun chapter3_tutorials example4
```

Now we have two options. First, we can record all the topics:

```
rosbag record -a
```

Or, second, record only some specific (user-defined) topics. In this case, it will make sense to record only the `example4` topics, so we will use the following:

```
rosbag record /temp /accel
```

By default, when we run the preceding command, the `rosbag` program subscribes to the node and starts recording the message in a bag file in the current directory with data as the name. Once you have finished the experiment or you want to stop recording, you only have to hit *Ctrl* + *C*. The following is an example of recording the data and the resulting bag file:

```
[ INFO] [1357815371.768263730]: Subscribing to /temp
[ INFO] [1357815371.771339658]: Subscribing to /accel
[ INFO] [1357815371.774950563]: Recording to 2013-01-10-10-56-11.bag.
```

You can see more options with `rosbag help record` that includes things such as the bag file size, the duration of the recording, and options to split the files into several ones of a given size. As we have mentioned before, the file can be compressed on the fly (using the `-j` option). In our honest opinion, this is only useful for small bandwidths because it also consumes some CPU time and might produce some message dropping. Also, we can increase the buffer (`-b`) size for the recorder in MB, which defaults to 256 MB, but it can be increased to some GB if the bandwidth is very high (especially with images).

It is also possible to include the call to `rosbag record` into a launch file. To do so, we must add a node like this:

```
<node pkg="rosbag" type="record" name="bag_record"
        args="/temp /accel"/>
```

Note that the topics and other arguments to the command are passed using the `args` argument. Also, it is important to say that when running from the `launch` file, the bag file is created by default in `~/.ros`, unless we give the name of the file with `-o` (prefix) or `-O` (full name).

Playing back a bag file

Now that we have a bag file recorded, we can use it to play back all the messages of the topics inside it. We need `roscore` running and nothing else. Then, we move to the folder with the bag file we want to play (there are two examples in the `bag` folder of this chapter's tutorials) and do this:

```
rosbag play 2013-01-10-10-56-11.bag
```

We will see the following output:

```
[ INFO] [1357820252.241049890]: Opening bag/2013-01-10-10-56-11.bag


Waiting 0.2 seconds after advertising topics... done.


Hit space to toggle paused, or 's' to step.
 [RUNNING]  Bag Time: 1357815375.147705   Duration: 2.300787 / 39.999868
```

In the terminal where we are playing the bag file, we can pause (hit Space bar) or move step by step (hit *S*), and, as usual, use *Ctrl* + *C* to finish it immediately. Once we reach the end of the file, it will close, but there is an option to loop (`-l`) that sometimes might be useful.

Automatically, we will see the topics with `rostopic list`:

```
/accel
/clock
/rosout
/rosout_agg
/temp
```

The `/clock` topic is part of the fact that we can instruct the system clock to simulate a faster playback. This can be configured using the `-r` option. In the `/clock` topic, the time for simulation at a configurable frequency with the `--hz` argument (it defaults to 100 Hz) is published.

Also, we could specify a subset of the topics in the file to be published. This is done with the `--topics` option. In order to see what we have inside the file, we would use `rosbag info <bag_file>`, which we will explain in the next section.

Inspecting all the topics and messages in a bag file using rxbag

There are two main ways to see what we have inside a bag file. The first one is very simple. We just type `rosbag info <bag_file>` and the result is something like this:

```
user@pc$ rosbag info bag/2013-01-10-10-56-11.bag
path:        bag/2013-01-10-10-56-11.bag
version:     2.0
duration:    40.0s
start:       Jan 10 2013 10:56:12.85 (1357815372.85)
end:         Jan 10 2013 10:56:52.85 (1357815412.85)
size:        10.9 KB
messages:    82
compression: none [1/1 chunks]
types:       geometry_msgs/Vector3 [4a842b65f413084dc2b10fb484ea7f17]
             std_msgs/Int32        [da5909fbe378aeaf85e547e830cc1bb7]
topics:      /accel    41 msgs    : geometry_msgs/Vector3
             /temp     41 msgs    : std_msgs/Int32user@pc$ rosbag info bag/2013-01-10-10-56-11.bag
```

We have information about the bag file itself, such as the creation date, duration, size, as well as the number of messages inside, and the compression (if any). Then, we have the list of data types inside the file, and finally the list of topics with their corresponding name, number of messages, and type.

The second way to inspect a bag file is extremely powerful. It is a graphical interface named **rxbag** that also allows playing back the files, viewing the images (if any), plotting scalar data, and also the raw structure of the messages. We only have to pass the name of the bag file, and we will see something like the following screenshot (for the previous bag file):

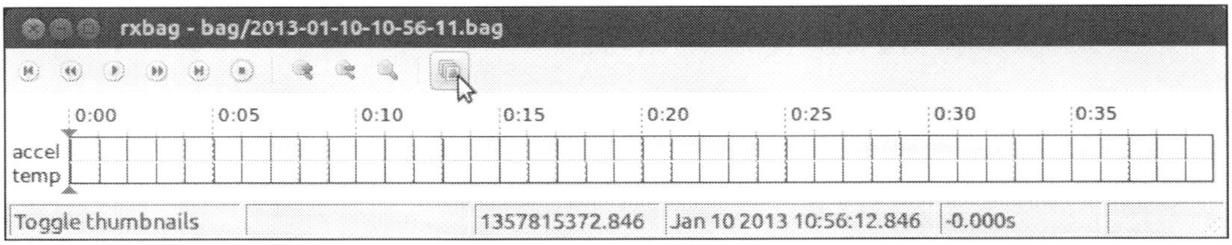

We have a timeline for all the topics where each message appears with a mark. In the case of images, we can enable the thumbnails to see them in the timeline (marked with the mouse pointer).

In the following screenshot, we can see how to access the **Raw**, **Plot**, and **Image** (if the topic is of the type Image) views for the topics in the file. This pop-up menu appears with a right-click over the timeline.

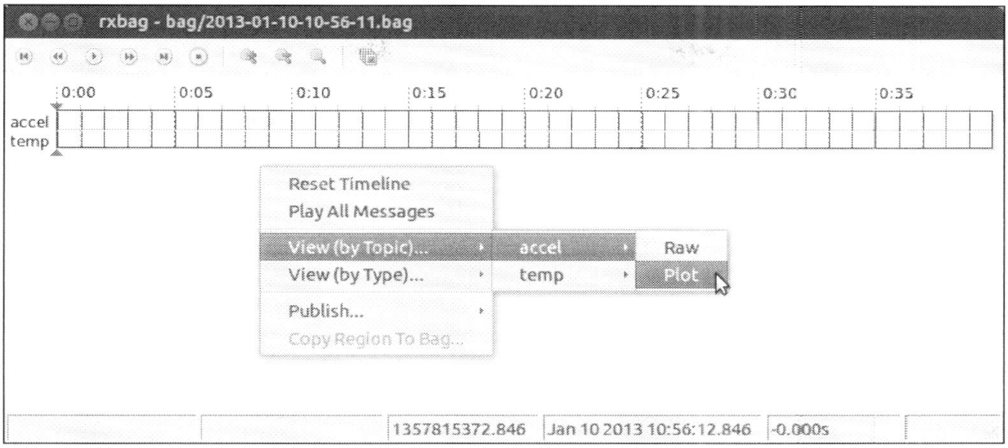

For `/accel`, we can plot all the fields in a single axis. To do so, once we are in the Plot view, we click on the gear button/icon and then select every field. Note that we can remove them later or create a different axis (in the bottom-right window). The plot is generated for all the values in the file, and a vertical line shows the current position in the playback.

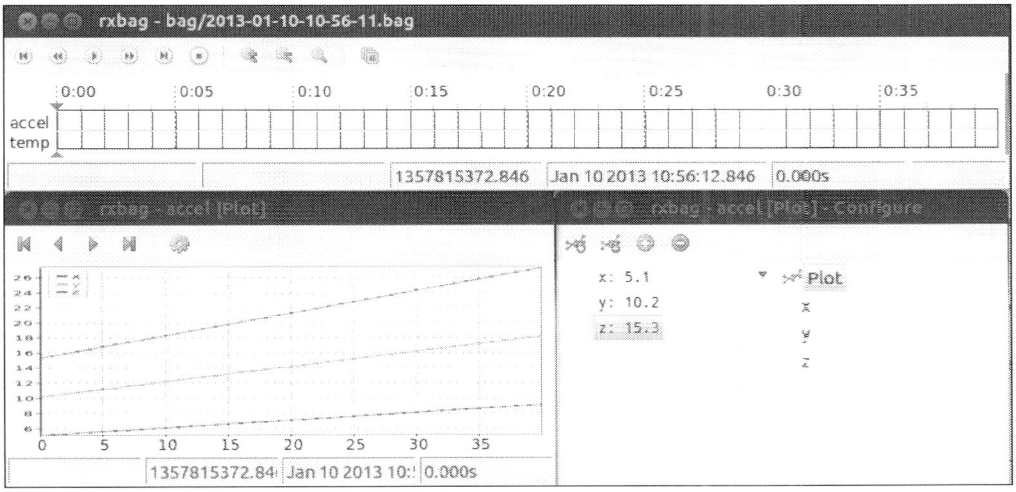

Note that we must have clicked on the **Play** button at least once to be able to plot the data. Then we can play, pause, stop, and move to the beginning or the end of the file.

The images are straightforward, and a simple window appears with the current frame with options to save them as image files in the disk.

rqt plugins versus rx applications

Since ROS Fuerte, the `rx` applications or tools are deprecated and we should instead use the `rqt` nodes. They are basically the same, only with a few of them incorporated with small updates, bug fixes, and new features. Also, they can be loaded as plugins into a single window/application, which is `rqt_gui`. We show the equivalent for the tools shown in this chapter in the following list:

- `rxconsole` is replaced by `rosrun rqt_console rqt_console`
- `rxgraph` is replaced by `rosrun rqt_graph rqt_graph`
- `rxplot` is replaced by `rosrun rqt_plot rqt_plot`
- `rxbag` is replaced by `rosrun rqt_bag rqt_bag`

Furthermore, being plugins that can also be run standalone, there exist more tools, such as a shell, a topic publisher, and a message type viewer. Even rviz has a plugin named `rqt_rviz` that can be integrated in the new `rqt_gui` interface; all this is fully integrated in ROS Groovy and Hydro where `rx` tools are deprecated but still in the bundle. The same happens for ROS Fuerte, which was the first release to incorporate the `rqt` tools.

Summary

After reading and running the code of this chapter, you will have learned to use many tools that will enable you to develop robotic systems faster, debug errors, and visualize your results so you can evaluate their quality or validate them. Some of the specific concepts and tools you will exploit the most in your life as a robotic developer are summarized as follows:

- Now you know how to include logging messages in your code with different levels of verbosity, which will help you debug errors in your nodes. For this purpose, you could also use the powerful tools included in ROS, such as the rxconsole interface. Additionally, you can also inspect or list the nodes running, the topics published, and the services provided in the whole system while running. This includes the inspection of the node graph using rxgraph.
- Regarding the visualization tools, you should be able to plot scalar data using `rxplot` for a more intuitive analysis of certain variables published by your nodes. Similarly, you can view more complex types (nonscalar ones). This includes images and 3D data using rviz.
- Finally, recording and playing back the messages of the topics available is now in your hands with `rosbag`. And you also know how to view the contents of a bag file with `rxbag`. This allows you to record the data from your experiments and process them later with your AI or robotics algorithms.

4
Using Sensors and Actuators with ROS

When you think of a robot, you would probably think of human-size robots with arms, a lot of sensors, and a wide field of locomotion systems.

Now that we know how to write small programs in ROS and manage them, we are going to work with sensors and actuators, something that can interact with the real world.

You can find a wide list of supported devices by ROS at `http://www.ros.org/wiki/Sensors`.

In this chapter, we will deal with the following:

- Cheap and common sensors for your projects
- 3D sensors as Kinect and laser rangefinders
- Using Arduino to connect more sensors or actuators

We know that it is impossible to explain all types of sensors in this chapter. For this reason, we have selected some of the most commonly used ones and those which are affordable for most users—regular, sporadic, or amateur.

Sensors and actuators can be organized in different categories: rangefinders, cameras, pose estimation devices, and so on. These will help you find the sensor or actuator that you are looking for more quickly.

Using a joystick or gamepad

I am sure that at one point or another you have used a joystick or a gamepad of a video console.

A joystick is nothing more than a series of buttons and potentiometers. With this device, you can perform or control a wide range of actions.

In ROS, a joystick is used to telecontrol a robot for changing its velocity or direction.

Before we start, we are going to install some packages. To install these packages in Ubuntu, execute the following:

```
$ sudo apt-get install ros-fuerte-joystick-drivers
$ sudo apt-get install ros-fuerte-joystick-drivers-tutorials
```

In these packages, you will find some code to learn how to use the joystick and a guide to create our packages.

First of all, connect your joystick to your computer. Now we are going to check if the joystick is recognized.

```
$ ls /dev/input/
```

We will see the following output:

```
by-id    event0  event2  event4  event6  event8  js0   mouse0
by-path  event1  event3  event5  event7  event9  mice
```

The port created is `js0`; we can check if it is working with the command `jstest`.

```
$ sudo jstest /dev/input/js0


Axes:  0:   0  1:   0  2:   0 Buttons:  0:off  1:off  2:off  3:off  4:off
5:off  6:off  7:off  8:off  9:off  10:off
```

Our joystick, **Logitech Attack 3**, has 3 axes and 11 buttons, and if we move the joystick, the values change.

Once you have checked the joystick, we are going to test it in ROS. To do this, you can use the `joy` and `joy_node` packages.

```
$ rosrun joy joy_node
```

If everything is OK, you will see this:

```
[ INFO] [1357571588.441808789]: Opened joystick: /dev/input/js0.
deadzone_: 0.050000.
```

How does joy_node send joystick movements?

With the `joy_node` package active, we are going to see the messages sent by this node. This will help us understand how it is sending the information of axes and buttons.

To see the messages sent by the node, we can use this command:

```
$ rostopic echo /joy
```

And then we can see each message sent.

```
---
header:
seq: 157
stamp:
  secs: 1357571648
  nsecs: 430257462
  frame_id: ''
axes: [-0.0, -0.0, 0.0]
buttons: [0, 0, 0, 0, 0, 0, 0, 0, 0, 0, 0]
---
```

You will see two main vectors, one for axes and another for buttons. Obviously, these vectors are used to publish the states of the buttons and axes of the real hardware.

If you want to know the message type, type the following command line in a shell:

```
$ rosnode type /joy
```

You will then obtain the type used by the message; in this case, it is `sensor_msgs/Joy`.

Now to see the fields used in the message, use the following command line:

```
$ rosmsg show sensor_msgs/Joy
```

And you will see this:

```
  uint32 seq
  time stamp
  string frame_id
float32[] axes
int32[] buttons
```

This is the message structure that you must use if you want to use a joystick with your developments. In the next section, you will learn how to write a node that subscribes to the joystick topic and generate moving commands to move the turtlesim.

Using joystick data to move a turtle in turtlesim

Now, we are going to create a node that gets data from `joy_node` and publishes topics to control turtlesim.

First, it is necessary to know the topic name where we will publish the messages. So, we are going to start turtlesim and do some investigations.

```
$ rosrun turtlesim turtlesim_node
```

To see the topic list, use the following command line:

```
$ rostopic list
```

You will then see the following output where `turtle1/command_velocity` is the topic we will use:

```
/rosout
/rosout_agg
/turtle1/color_sensor
/turtle1/command_velocity
/turtle1/pose
```

Now we need to know the topic type. Use the following command line to see it:

```
$ rostopic type /turtle1/command_velocity
```

You will see this output:

```
turtlesim/Velocity
```

To know the content of this message, execute the following command line:

```
$ rosmsg show turtlesim/Velocity
```

You will then see the two fields, which are used to send the velocity.

```
float32 linear
float32 angular
```

OK, now that we have localized the topic and the structure to use, it is time to create a program to generate velocity commands using data from the joystick.

Create a new file, `example.1.cpp`, in the `chapter4_tutorials/src` directory and type in the following code snippet:

```
#include<ros/ros.h>
#include<turtlesim/Velocity.h>
#include<sensor_msgs/Joy.h>
#include<iostream>

using namespace std;

class TeleopJoy{
public:
  TeleopJoy();
private:
  void callBack(const sensor_msgs::Joy::ConstPtr& joy);
  ros::NodeHandle n;
  ros::Publisher pub;
  ros::Subscriber sub;
  int i_velLinear, i_velAngular;
};

TeleopJoy::TeleopJoy()
{
n.param("axis_linear",i_velLinear,i_velLinear);
n.param("axis_angular",i_velAngular,i_velAngular);
```

```
pub = n.advertise<turtlesim::Velocity>("turtle1/command_velocity",1);
sub = n.subscribe<sensor_msgs::Joy>("joy", 10, &TeleopJoy::callBack,
this);
}

void TeleopJoy::callBack(const sensor_msgs::Joy::ConstPtr& joy)
{
turtlesim::Velocity vel;
vel.angular = joy->axes[i_velAngular];
vel.linear = joy->axes[i_velLinear];
pub.publish(vel);
}

int main(int argc, char** argv)
{
ros::init(argc, argv, "teleopJoy");
    TeleopJoy teleop_turtle;

ros::spin();
}
```

Now we are going to break the code to explain how it works.

In the `main` function, we create an instance of the class `TeleopJoy`.

```
int main(int argc, char** argv)
{
   ...
   TeleopJoy teleop_turtle;
  ...
```

In the constructor, four variables are initialized. The first two variables are filled using data from the Parameter Server. These variables are joystick axes. The next two variables are the publisher and subscriber. The publisher will publish a topic with the type `turtlesim::Velocity`; this type is declared in the `turtlesim` package. The subscriber will get data from the topic with the name `joy`. The node that is handling the joystick sends this topic.

```
TeleopJoy::TeleopJoy()
{
n.param("axis_linear",i_velLinear,i_velLinear);
n.param("axis_angular",i_velAngular,i_velAngular);

pub = n.advertise<turtlesim::Velocity>("command_velocity",1);
sub = n.subscribe<sensor_msgs::Joy>("joy", 10, &TeleopJoy::callBack,
this);

}
```

Each time the node receives a message, the function `callBack` is called. We create a new variable with the name `vel`, which will be used to publish data. The values of the axes of the joystick are assigned to the `vel` variable. In this part, you can make some process with the data received before publishing it.

```
void TeleopJoy::callBack(const sensor_msgs::Joy::ConstPtr& joy)
{
turtlesim::Velocity vel;
vel.angular = joy->axes[i_velAngular];
vel.linear = joy->axes[i_velLinear];
pub.publish(vel);
}
```

Finally, the topic is published using `pub.publish(vel)`.

We are going to create a `launch` file for this example. In the `launch` file, we declare some data for the Parameter Server and launch the `joy` and `example1` nodes.

Copy the following code step to a new file, `example1.launch`, in the `chapter4_tutorials/src` directory:

```
<launch>

<node pkg="turtlesim" type="turtlesim_node" name="sim"/>
<node pkg="chapter4_tutorials" type="example1" name="example1" />
<param name="axis_linear" value="1" type="int" />
<param name="axis_angular" value="0" type="int" />

<node respawn="true" pkg="joy"type="joy" name="teleopJoy">
    <param name="dev" type="string" value="/dev/input/js0" />
    <param name="deadzone" value="0.12" />
</node>

</launch>
```

You will notice that in the `launch` file, there are three different nodes: `example1`, `sim`, and `joy`.

There are four parameters in the `launch` file; these parameters will add data to the Parameter Server, and it will be used by our node. The `axis_linear` and `axis_angular` parameters will be used to configure the axis of the joystick. If you want to change the axis configuration, you only need to change the value and put the number of the axes you want to use. The `dev` and `deadzone` parameters will be used to configure the port where the joystick is connected, and the dead zone is the region of movement that is not recognized by the device.

To run the `launch` file, use the following command line:

```
$ roslaunch chapter4_tutorials example1.launch
```

You can see if everything is fine by checking the running nodes and the topic list using `rosnode list` and `rostopic list`. If you want to see it graphically, use `rxgraph`.

Using a laser rangefinder – Hokuyo URG-04lx

In mobile robotics, it is very important to know where the obstacles are, the outline of a room, and so on. The robots use maps to navigate and move across unknown spaces. The sensor used for these purposes is **Lidar**. This sensor is used to measure distances between the robot and objects.

In this section, you will learn how to use a low-cost Lidar that is widely used in robotics. This sensor is the **Hokuyo URG-04lx** rangefinder. You can obtain more information in the following link: `http://www.hokuyo-aut.jp/`. The Hokuyo rangefinder is a device used for navigation and building maps in real time.

In this section, you will learn how to use it using the standard drivers in ROS and how to modify the data.

The model Hokuyo URG-04lx is a low-cost rangefinder commonly used in robotics. It has a very good resolution and is very easy to use.

To start with, we are going to install the drivers for the laser.

```
$ sudo apt-get install ros-fuerte-laser-drivers
```

Once installed, we are going to check if everything is OK. Connect your laser and check if the system can detect it and if it is configured correctly.

```
$ ls -l /dev/ttyACM0
```

When the laser is connected, the system can see it using the following command line:

```
crw-rw---- 1 root dialout 166, 0 Jan 13 11:09 /dev/ttyACM0
```

In our case, we need to reconfigure the laser device to give ROS the access to use it; that is, we need to give appropriate permissions.

```
$ sudo chmod a+rw /dev/ttyACM0
```

You will then see the following output:

```
crw-rw-rw- 1 root dialout 166, 0 Jan 13 11:09 /dev/ttyACM0
```

Once everything is OK, we are going to switch on the laser. Start `roscore` in one shell and in another shell execute the following command:

```
$ rosrun hokuyo_node hokuyo_node
```

If everything is fine, you will see the following output:

```
[ INFO] [1358076340.184643618]: Connected to device with ID: H1000484
```

Understanding how the laser sends data in ROS

To check if the node is sending data, use `rostopic`.

```
$ rostopic list
```

And you will see the following topics:

```
/diagnostics
/hokuyo_node/parameter_descriptions
/hokuyo_node/parameter_updates
/rosout
/rosout_agg
/scan
```

The topic `/scan` is the topic where the node is publishing. The type of data used by the node is shown as follows:

```
$ rostopic type /scan
```

You will then see the message type used to send information of the laser.

```
sensor_msgs/LaserScan
```

You can see the structure of the message using `$ rosmsg show sensor_msgs/LaserScan`.

To learn a little bit more on how the laser works and what data it is sending, we are going to use the `rostopic` command to see a real message.

```
$ rostopic echo /scan
```

Then you will see the following message sent by the laser:

```
---
header:
seq: 3895
stamp:
  secs: 1358076731
  nsecs: 284896750
  frame_id: laser
...
ranges: [1.1119999885559082, 1.1119999885559082, 1.1109999418258667, ...]
intensities: []
---
```

These data are difficult to understand for humans. If you want to see the data in a more friendly and graphical way, it is possible do it using rviz. Type the following command line in a shell to launch rviz with the correct configuration file:

```
$ rosrun rviz rviz -d 'rospack find hokuyo_node'/hokuyo_test.vcg
```

Downloading the example code

You can download the example code files for all Packt books you have purchased from your account at `http://www.packtpub.com`. If you purchased this book elsewhere, you can visit `http://www.packtpub.com/support` and register to have the files e-mailed directly to you. You can also download these code files from `https://github.com/AaronMR/Learning_ROS_for_Robotics_Programming`.

The following screenshot shows a graphical representation of the message:

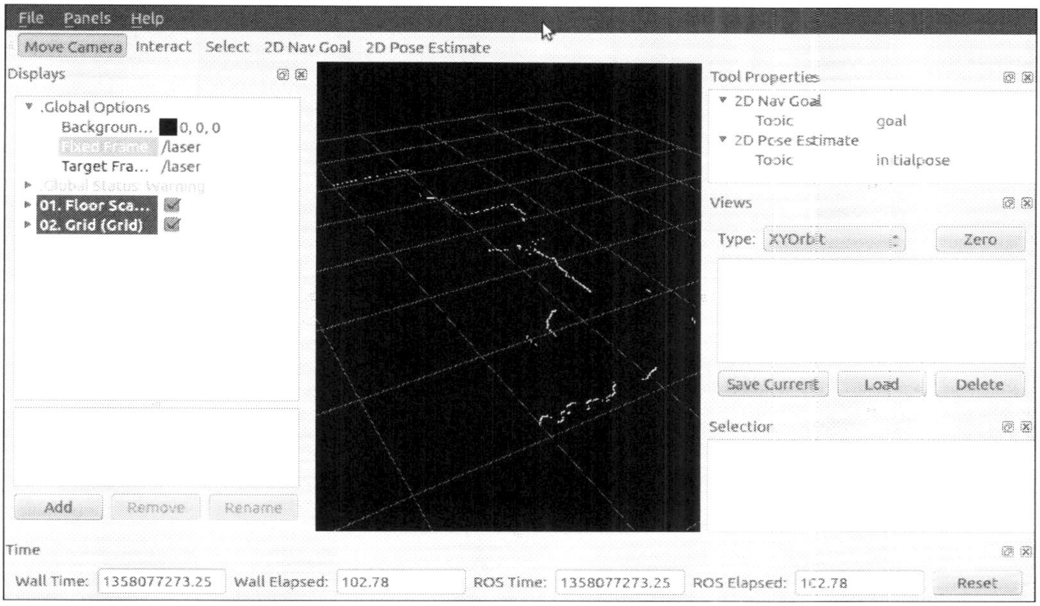

You will see the contour on the screen. If you move the laser sensor, you will see the contour changing.

Accessing the laser data and modifying it

Now, we are going to make a node get the laser data, do something with it, and publish the new data. Perhaps, this will be useful sometime, and with this example, you will learn how to do it.

Copy the following code snippet to the `example2.cpp` file in your `/chapter4_tutorials/src` directory:

```
#include <ros/ros.h>
#include "std_msgs/String.h"
#include <sensor_msgs/LaserScan.h>
#include<stdio.h>

using namespace std;
class Scan2{
public:
```

```
    Scan2();
  private:
    ros::NodeHandle n;
      ros::Publisher scan_pub;
      ros::Subscriber scan_sub;
    void scanCallBack(const sensor_msgs::LaserScan::ConstPtr& scan2);
  };

  Scan2::Scan2()
  {
  scan_pub = n.advertise<sensor_msgs::LaserScan>("/scan2",1);
  scan_sub = n.subscribe<sensor_msgs::LaserScan>("/scan",1,
  &Scan2::scanCallBack, this);
  }

  void Scan2::scanCallBack(const sensor_msgs::LaserScan::ConstPtr&
  scan2)
  {
  int ranges = scan2->ranges.size();
      //populate the LaserScan message
  sensor_msgs::LaserScan scan;
      scan.header.stamp = scan2->header.stamp;
      scan.header.frame_id = scan2->header.frame_id;
      scan.angle_min = scan2->angle_min;
      scan.angle_max = scan2->angle_max;
      scan.angle_increment = scan2->angle_increment;
      scan.time_increment = scan2->time_increment;
  scan.range_min = 0.0;
      scan.range_max = 100.0;
  scan.ranges.resize(ranges);

  for(int i = 0; i < ranges; ++i){
      scan.ranges[i] = scan2->ranges[i] + 1;
      }
  scan_pub.publish(scan);
  }

  int main(int argc, char** argv){
  ros::init(argc, argv, "example2_laser_scan_publisher");
      Scan2 scan2;
  ros::spin();
  }
```

We are going to break the code and see what it is doing.

In the `main` function, we initialize the node with the name `example2_laser_scan_publisher`, and create an instance of the class that we have created in the file.

In the constructor, we will create two topics: one of them will subscribe to the other topic, which is the original data from the laser. The second topic will publish the new modified data from the laser.

This example is very simple; we are only going to add the `1` unit to data received from the laser topic and publish it again. We do that in the `scanCallBack()` function. Take the input message and copy all the fields to another variable and take the field where the data is stored and add the `1` unit. Once the new value is stored, publish the new topic.

```
void Scan2::scanCallBack(const sensor_msgs::LaserScan::ConstPtr&
scan2)
{
    ...
sensor_msgs::LaserScan scan;
    scan.header.stamp = scan2->header.stamp;
 ...
 ...
    scan.range_max = 100.0;
scan.ranges.resize(ranges);

for(int i = 0; i < ranges; ++i){
    scan.ranges[i] = scan2->ranges[i] + 1;
    }

scan_pub.publish(scan);
}
```

Creating a launch file

To launch everything, we are going to create a launch file, `example2.launch`.

```
<launch>
  <node pkg="hokuyo_node" type="hokuyo_node" name="hokuyo_node"/>
  <node pkg="rviz" type="rviz" name="rviz"
  args="-d $(find chapter4_tutorials)/example2.vcg"/>

  <node pkg="chapter4_tutorials" type="example2" name="example2" />
</launch>
```

Now if you launch the `example2.launch` file, three nodes will start: `hokuyo_node`, `rviz`, and `example2`. You will see the rviz screen with the two-lasers contour. The green contour is the new data.

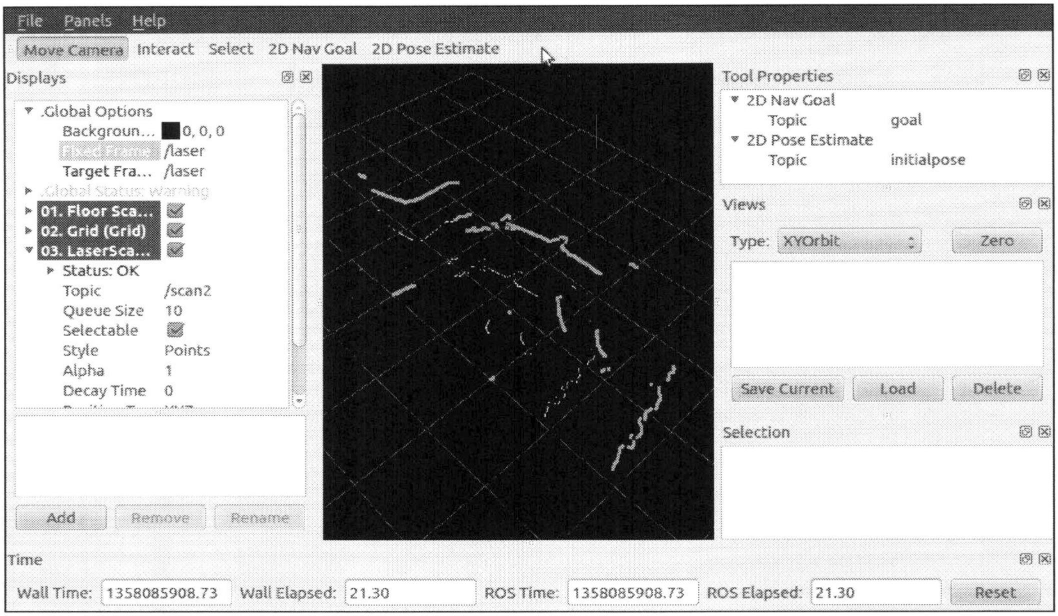

Using the Kinect sensor to view in 3D

The Kinect sensor is a flat black box that sits on a small platform when placed on a table or shelf near the television you're using with your Xbox 360. This device has the following three sensors that we can use for vision and robotics tasks:

- A color VGA video camera to see the world in color
- The depth sensor, which is an infrared projector and a monochrome CMOS sensor working together, to see objects in 3D
- A multiarray microphone that is used to isolate the voices of the players from the noise in the room

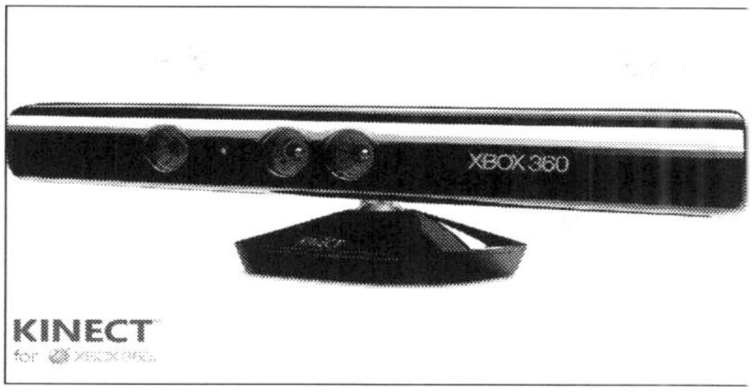

In ROS, we are going to use two of these sensors: the RGB camera and the depth sensor. In the latest version of ROS, you can even use three.

Before we start using it, we need to install the packages and drivers. Use the following command lines to install it:

```
$ sudo apt-get install ros-fuerte-openni-camera ros-fuerte-openni-launch
$ rosstack profile
$ rospackage profile
```

Once installed, plug the Kinect sensor and we will run the nodes to start using it. In a shell, start `roscore`. In another shell run the following command lines:

```
$ rosrun openni_camera openni_node
$ openni_launch openni.launch
```

If everything goes fine, you will not see any error messages.

How does Kinect send data from the sensors and how to see it?

Now we are going to see what we can do with these nodes. List the topics that you have created using this command:

```
$ rostopic list
```

Then, you will see a lot of topics, but the most important ones for us are the following:

```
...
/camera/rgb/image_color
/camera/rgb/image_mono
/camera/rgb/image_raw
/camera/rgb/image_rect
/camera/rgb/image_rect_color
...
```

We will see a lot of topics created by nodes. If you want to see one of the sensors, the RGB camera for example, you can use the topic `/camera/rgb/image_color`. To see the image from the sensor, we are going to use the `image_view` package. Type the following command in a shell:

```
$ rosrun image_view image_view image:=/camera/rgb/image_color
```

Note that we need to rename (remap) the image topic to `/camera/rgb/image_color`. If everything is fine, a new window appears showing the image from Kinect.

If you want to see the depth sensor, you can do the same just by changing the topic in the last command line:

```
$ rosrun image_view image_view image:=/camera/depth/image
```

You will then see an image similar to the following output:

Another important topic is the one that sends the point cloud data. This kind of data is a 3D representation of the depth image. You can find this data in the following topics: `/camera/depth/points`, `/camera/depth_registered/points` and so on.

We are going to see what type of message this is. To do this, use `rostopic type`. To see the fields of a message, we can use `rostopic type /topic_name | rosmsg show`. In this case, we are going to use the `/camera/depth/points` topic.

```
$ rostopic type /camera/depth/points | rosmsg show
```

To see the official specification of the message, visit `http://ros.org/doc/api/sensor_msgs/html/msg/PointCloud2.html`.

If you want to visualize this type of data, run rviz in a new shell and add a new **PointCloud2** data visualization. Depending on your computer, you will see a 3D image in real time; if you move in front of the sensor, you will see yourself moving in 3D as you can see in the following screenshot:

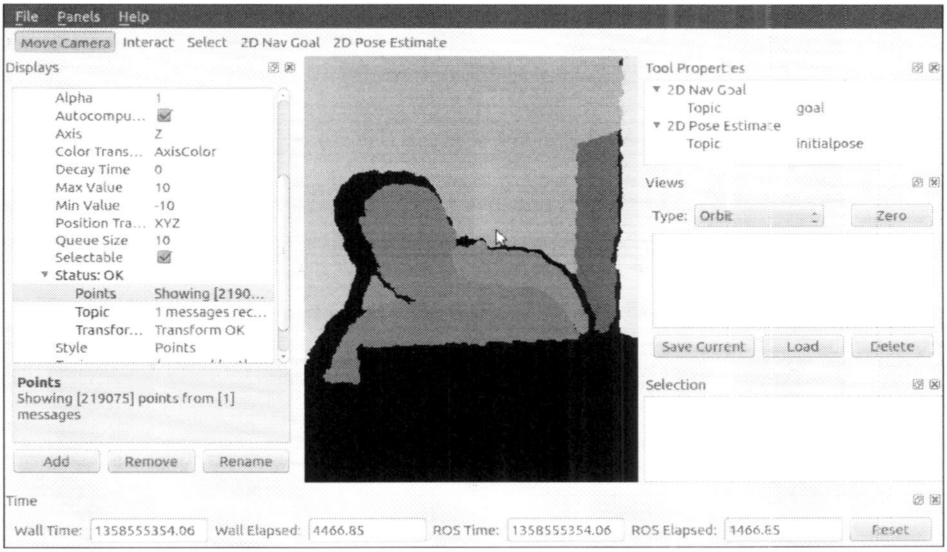

Creating an example to use Kinect

Now we are going to implement a program to generate a node that filters the point cloud from the Kinect sensor. This node will apply a filter to reduce the number of points on the original data. It will make a down sampling of the data.

Create a new file, `example3.cpp`, in your `chapter4_tutorials/src` directory and type in the following code snippet:

```
#include <ros/ros.h>
#include <sensor_msgs/PointCloud2.h>
#include <pcl/ros/conversions.h>
#include <pcl/point_cloud.h>
#include <pcl/point_types.h>
#include <pcl/filters/voxel_grid.h>
```

```
ros::Publisher pub;
void cloud_cb (const sensor_msgs::PointCloud2ConstPtr& input)
{
  // ... do data processing
sensor_msgs::PointCloud2 output;
pcl::VoxelGrid<sensor_msgs::PointCloud2> sor;
sor.setInputCloud(input);
sor.setLeafSize(0.02f,0.02f,0.02f);
sor.filter(output);

  // Publish the data
  pub.publish (output);
}

intmain (int argc, char** argv)
{
  // Initialize ROS
  ros::init (argc, argv, "my_pcl_tutorial");
ros::NodeHandle nh;

  // Create a ROS subscriber for the input point cloud
ros::Subscriber sub = nh.subscribe ("/camera/depth/points", 1, cloud_
cb);

  // Create a ROS publisher for the output point cloud
pub = nh.advertise<sensor_msgs::PointCloud2> ("output", 1);

  // Spin
  ros::spin ();
}
```

This sample is based on the tutorial of **Point Cloud Library (PCL)**. You can see it at `http://pointclouds.org/documentation/tutorials/voxel_grid.php#voxelgrid`.

All the work is done in the `cb()` function. This function is called when a message arrives. We create a variable `sor` with the `VoxelGrid` type, and the range of the grid is changed in `sor.setLeafSize()`. These values will change the grid used for the filter. If you increment the value, you will obtain less resolution and less points on the point cloud.

```
cloud_cb (const sensor_msgs::PointCloud2ConstPtr& input)
{
...
  pcl::VoxelGrid<sensor_msgs::PointCloud2> sor;
```

```
    ...
    sor.setLeafSize(0.02f,0.02f,0.02f);
  ...
  }
```

If we open rviz now with the new node running, we will see the new point cloud on the window, and you will directly notice that the resolution is less than the original data.

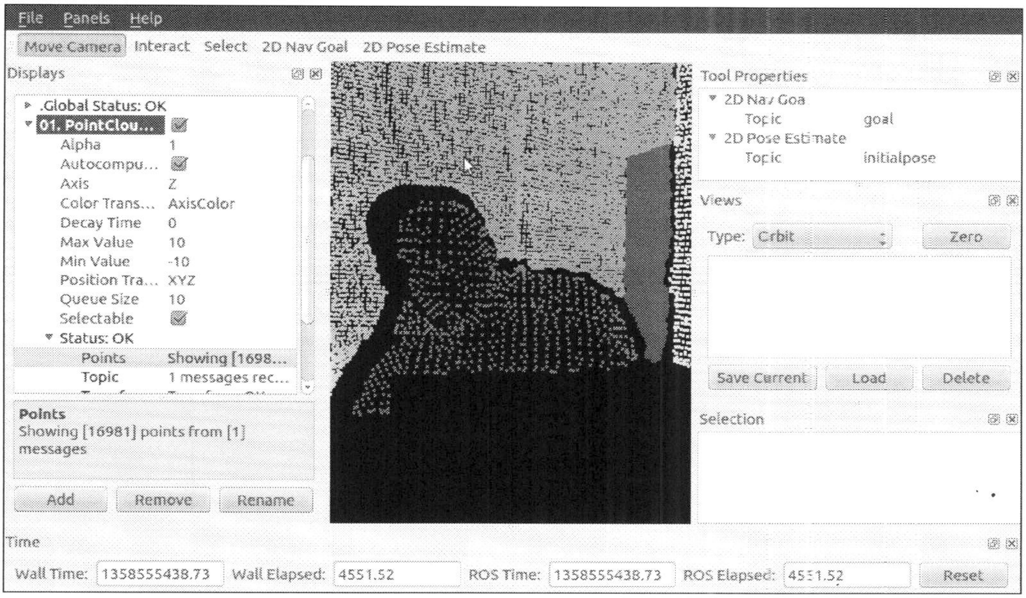

On rviz, you can see the number of points that a message has. For original data, we can see that the number of points is 2,19,075. With the new point cloud, we obtain a number of points of 16,981. As you can see, it is a huge reduction of data.

At `http://pointclouds.org/`, you will find more filters and tutorials on how to use this kind of data.

Using servomotors – Dynamixel

In mobile robots, servomotors are widely used. These kind of actuators are used to move sensors, wheels, or robotic arms. A low-cost solution is to use RC servomotors. It provides a movement range of 180 degrees and a high torque for the existing servomotors.

The servomotor that we will explain in this section is a new type of servomotor designed and used for robotics. This is the **Dynamixel** servomotor.

Dynamixel is a line-up, high-performance networked actuator for robots developed by ROBOTIS, a Korean manufacturer. ROBOTIS is also the developer and manufacturer for OLLO, Bioloid, and DARwIn-OP DXL. These are used by numerous companies, universities, and hobbyists due to their versatile expansion capability, powerful feedback functions, position, speed, internal temperature, input voltage, and other features, and its simple daisy chain topology for simplified wiring connections.

In the following image, you can see the Dynamixel AX-12 and the USB interface. Both are used in this example.

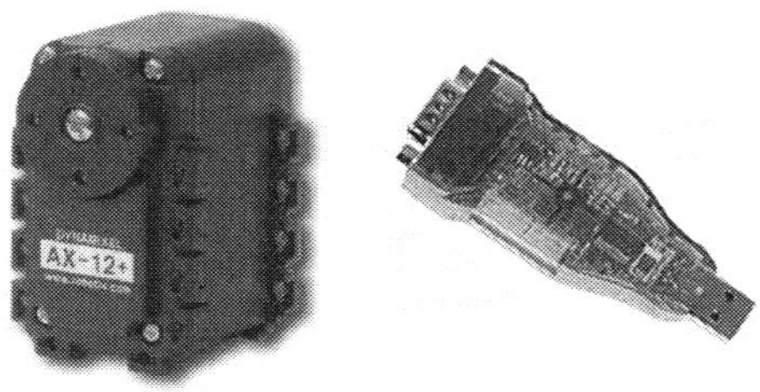

First, we are going to install the necessary packages and drivers. Type the following command line in a shell:

```
$ sudo apt-get install ros-fuerte-dynamixel-motor
```

Once installed, connect the dongle to the computer and check if it is detected. Normally, it will create a new port with the name ttyUSBX inside your `/dev/` folder. If you see this port, everything is OK, and now we can start the nodes to play a little with the servomotor.

In a shell, start `roscore` and in another shell, type the following command line:

```
$ roslaunch dynamixel_tutorials controller_manager.launch
```

If the motor or motors are connected, you will see the motors detected by the driver. In our case, a motor with the ID 6 is detected and configured.

```
process[dynamixel_manager-1]: started with pid [3966]
[INFO] [WallTime: 1359377042.681841] pan_tilt_port: Pinging motor IDs 1 through 25...
[INFO] [WallTime: 1359377044.846779] pan_tilt_port: Found 1 motors - 1 AX-12 [6], initialization complete.
```

How does Dynamixel send and receive commands for the movements?

Once you have launched the `controller_manager.launch` file, you will see a list of topics. Remember to use the following command line to see these topics:

```
$ rostopic list
```

These topics will show the state of the motors configured.

```
/diagnostics
/motor_states/pan_tilt_port
/rosout
/rosout_agg
```

If you see `/motor_states/pan_tilt_port` with the `rostopic echo` command, you will see the state of all motors, in our case, only the motor with the ID 6; however, we cannot move the motors with these topics, so we need to run the next `launch` file to do it.

This `launch` file will create the necessary topics to move the motors.

```
$ roslaunch dynamixel_tutorials controller_spawner.launch
```

The topic list will have two new topics added to the list. One of the new topics will be used to move the servomotor as follows:

```
/diagnostics
/motor_states/pan_tilt_port
/rosout
/rosout_agg
/tilt_controller/command
/tilt_controller/state
```

To move the motor, we are going to use `/tilt_controller/command` that will publish a topic with the `rostopic pub` command. First, you need to see the fields of the topic and the type. To do that, use the following command lines:

```
$ rostopic type /tilt_controller/command

std_msgs/Float64
```

As you can see, it is a `Float64` variable. This variable is used to move the motor to a position measured in radians. So, to publish a topic, use the following:

```
$ rostopic pub /tilt_controller/command std_msgs/Float64 -- 0.5
```

Once the command is executed, you will see the motor moving and it will stop at 0.5 radians or 28.6478898 degrees.

Creating an example to use the servomotor

Now, we are going to show you how to move the motor using a node. Create a new file, `example4.cpp`, in your `/chapter4_tutorials/src` directory with the following code snippet:

```
#include<ros/ros.h>
#include<std_msgs/Float64.h>
#include<stdio.h>

using namespace std;

class Dynamixel{
  private:
    ros::NodeHandle n;
    ros::Publisher pub_n;
  public:
    Dynamixel();
    int moveMotor(double position);
};

Dynamixel::Dynamixel(){
  pub_n = n.advertise<std_msgs::Float64>("/tilt_controller/
command",1);
}
int Dynamixel::moveMotor(double position)
{
  std_msgs::Float64 aux;
  aux.data = position;
  pub_n.publish(aux);
  return 1;
}

int main(int argc,char** argv)
{
  ros::init(argc, argv, "example4_move_motor");
  Dynamixel motors;

  float counter = -180;
  ros::Rate loop_rate(100);
```

```
    while(ros::ok())
    {
      if(counter < 180)
      {
        motors.moveMotor(counter*3.14/180);
        counter++;
      }else{
        counter = -180;
      }
      loop_rate.sleep();
    }
  }
```

This node will move the motor continuously from -180 to 180 degrees. It is a simple example, but you can use it to make complex movements or control more motors. We assume that you understand the code and it is not necessary to explain it. Note that you are publishing data to the `/tilt_controller/command` topic; this is the name of the motor.

Using Arduino to add more sensors and actuators

Arduino is an open source electronics prototyping platform based on flexible, easy-to-use hardware and software. It's intended for artists, designers, hobbyists, and anyone interested in creating interactive objects or environments.

ROS can use this type of device with the package `rosserial`. Basically, Arduino is connected to the computer using a serial connection and data is transmitted using this port. With `rosserial`, you can also use a lot of devices controlled by a serial connection; for example, GPS, servo controllers, and so on.

First, we need to install the package. To do this, we use the following command lines:

```
$ sudo apt-get install ros-fuerte-rosserial
$ rosstack profile
$ rospackage profile
```

OK, we assume that you have the Arduino IDE installed. If not, just follow the steps described at `http://arduino.cc/en/Main/Software`.

Once you have the package and the IDE installed, it is necessary to copy `ros_lib` from the `rosserial` package to the `sketchbook` library folder, which is created on your computer after running the Arduino IDE.

```
$ roscd rosserial_arduino/libraries
$ cp -r ros_lib <sketchbook>/libraries/ros_lib
```

Creating an example to use Arduino

Now, we are going to upload an example program from the IDE to Arduino. Select the Hello World sample and upload the sketch.

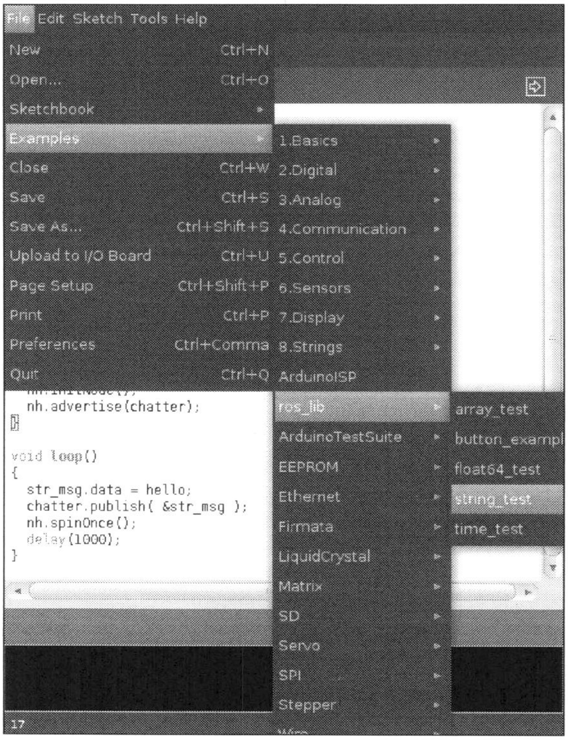

The code is very similar to the following code. In this code, you can see an `include` line with the `ros.h` library. This library is the `rosserial` library, which we have installed before. Also, you can see a library with the message to send with a topic; in this case, the `std_msgs/String` type.

The following code snippet is present in the `example5_1.ino` file:

```
#include <ros.h>
#include <std_msgs/String.h>

ros::NodeHandle  nh;

std_msgs::String str_msg;
ros::Publisher chatter("chatter", &str_msg);

char hello[19] = "chapter4_tutorials";

void setup()
{
  nh.initNode();
  nh.advertise(chatter);
}

void loop()
{
  str_msg.data = hello;
  chatter.publish( &str_msg );
  nh.spinOnce();
delay(1000);
}
```

In the `setup()` function, the name of the topic is set, in this case, it is called `chatter`. Now, we need to start a node to hear the port and publish the topics sent by Arduino on the ROS network. Type the following command in a shell. Remember to run `roscore`.

```
$ rosrun rosserial_python serial_node.py /dev/ttyUSB0
```

Now, you can see the messages sent by Arduino with the `rostopic echo` command.

```
$ rostopic echo chatter
```

You will see the following data in the shell:

```
data: chapter4_tutorials
```

The last example is about the data sent from Arduino to the computer. Now, we are going to use an example where Arduino will subscribe to a topic and will change the LED state connected to the pin number 13. The name of the example that we are going to use is `blink`; you can find this in the Arduino IDE by navigating to **File | Examples | ros_lib | Blink**.

The following code snippet is present in the `example5_2.ino` file:

```
#include <ros.h>
#include <std_msgs/Empty.h>

ros::NodeHandle nh;
void messageCb( const std_msgs::Empty& toggle_msg){
digitalWrite(13, HIGH-digitalRead(13));   // blink the led
}

ros::Subscriber<std_msgs::Empty> sub("toggle_led", &messageCb );

void setup()
{
  pinMode(13, OUTPUT);
  nh.initNode();
  nh.subscribe(sub);
}

void loop()
{
  nh.spinOnce();
  delay(1);
}
```

Remember to launch the node to communicate with the Arduino board.

```
$ rosrun rosserial_python serial_node.py /dev/ttyUSB0
```

Now if you want to change the LED status, you can use the `rostopic pub` command to publish the new state.

```
$ rostopic pub toggle_led std_msgs/Empty -once


publishing and latching message for 3.0 seconds
```

You will notice that the LED has changed its status; if the LED was on, it will now turn off. To change the status again, you only have to publish the topic once more.

```
$ rostopic pub toggle_led std_msgs/Empty --once

publishing and latching message for 3.0 seconds
```

Now you can use all the devices available for Arduino on ROS. This is very useful because you have access to cheap sensors and actuators to implement your robots.

> When we were writing the chapter, we noticed that some Arduino do not work with `rosserial`; for instance, Arduino Leonardo. So, be careful with the selection of the device to use with this package.
>
> We didn't face any problems while working with Arduino Mega, Arduino Duemilanove, and Arduino Nano.

Using the IMU – Xsens MTi

> *An inertial measurement unit, or IMU, is an electronic device that measures and reports on a craft's velocity, orientation, and gravitational forces, using a combination of accelerometers and gyroscopes, sometimes also magnetometers. IMUs are typically used to maneuver aircraft, including unmanned aerial vehicles (UAVs), among many others, and spacecraft, including satellites and landers.*
>
> *– Source: Wikipedia*

In the following image, you can see the Xsens MTi, which is the sensor used in this section.

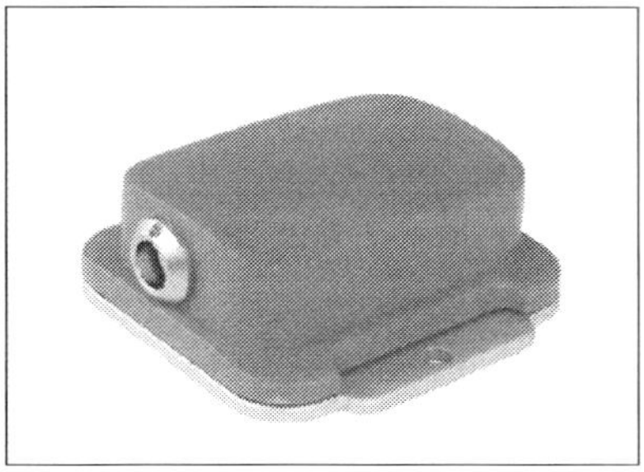

In this section, we will learn how to use it in ROS, the topics published by the sensor, and a small example with code to take data from the sensor and publish a new topic.

You can use a lot of IMU devices with ROS. In this section, we will use the Xsens IMU, which is necessary to install the right drivers. But if you want to use MicroStrain 3DM-GX2 or Wiimote with Wii Motion Plus, you need to download the following drivers:

- MicroStrain 3DM-GX2 IMU is available at `http://www.ros.org/wiki/microstrain_3dmgx2_imu`
- Wiimote with Wii Motion Plus is available at `http://www.ros.org/wiki/wiimote`

To use our device, we are going to use the `lse_xsens_mti` driver. You can install it using:

```
$ svn co http://isr-uc-ros-pkg.googlecode.com/svn/stacks/lse_imu_drivers/trunk/
```

We also need to install two packages because the driver depends on them.

```
$ svn co http://isr-uc-ros-pkg.googlecode.com/svn/stacks/serial_communication/trunk/cereal_port/
$ sudo apt-get install ros-fuerte-gps_umd
```

Recall installing them inside `ROS_PACKAGE_PATH`. Once installed, refresh the stacks and package installation.

```
$ rosstack profile
$ rospack profile
```

Now, we are going to start the IMU and see how it works. In a shell, start `roscore` and in another shell run the following command:

```
$ rosrun lse_imu_drivers mti_node
```

Make sure that the device is connected to the `/dev/ttyUSB0` port; if not, you must change the port by changing the code or using the Parameter Server as we have seen in other chapters.

How does Xsens send data in ROS?

If everything is fine, you can see the topic list using the `rostopic` command.

```
$ rostopic list
```

The node will publish three topics. We will work with `/imu/data` in this section. First of all, we are going to see the type and data sent by this topic. To see the type and the fields, use the following command lines:

```
$ rostopic type /imu/data
$ rostopic type /imu/data | rosmsg show
```

The fields are used to indicate the orientation, acceleration, and velocity. In our example, we will use the `orientation` field. Check a message to see a real example of the data sent. You can do it with the following command:

```
$ rostopic echo /imu/data

---
header:
seq: 288
stamp:
  secs: 1330631562
  nsecs: 789304161
frame_id: xsens_mti_imu
orientation:
x: 0.00401890464127
y: -0.00402884092182
z: 0.679586052895
w: 0.73357373476
...
---
```

If you observe the orientation field, you will see four variables instead of three, as you would probably expect. This is because in ROS, the spatial orientation is represented using quaternions. You can find a lot of literature on this concise and nonambiguous orientation representation on the Internet.

Creating an example to use Xsens

Now that we know the type of data sent and what data we are going to use, let's start with the example.

In this example, we are going to use the IMU to move the turtlesim. To do this, we need to take the data from the quaternion and convert it to Euler angles (roll, pitch, and yaw) and also take the rotation values around the x and y axes (roll and pitch) to move the turtle with a linear and angular movement.

The following code snippet is similar to the joystick code. Create a new file, `example6.cpp`, in your `chapter4_tutorials/src/` directory and type in the following code:

```
#include<ros/ros.h>
#include<turtlesim/Velocity.h>
#include<sensor_msgs/Imu.h>
#include<iostream>
#include<tf/LinearMath/Matrix3x3.h>
#include<tf/LinearMath/Quaternion.h>

using namespace std;

class TeleopImu{
public:
  TeleopImu();
private:
  void callBack(const sensor_msgs::Imu::ConstPtr& imu);
  ros::NodeHandle n;
  ros::Publisher pub;
  ros::Subscriber sub;
};

TeleopImu::TeleopImu()
{
  pub = n.advertise<turtlesim::Velocity>("turtle1/command_velocity",1);
  sub = n.subscribe<sensor_msgs::Imu>("imu/data", 10,
  &TeleopImu::callBack, this);
}

void TeleopImu::callBack(const sensor_msgs::Imu::ConstPtr& imu)
{
  turtlesim::Velocity vel;
  tf::Quaternion bq(imu->orientation.x,imu->orientation.y,imu-
  >orientation.z,imu->orientation.w);
    double roll,pitch,yaw;
    tf::Matrix3x3(bq).getRPY(roll,pitch,yaw);
      vel.angular = roll;
      vel.linear = pitch;
  pub.publish(vel);

}

int main(int argc, char** argv)
{
ros::init(argc, argv, "teleopImu");
    TeleopImu teleop_turtle;

ros::spin();
}
```

The node will subscribe to the topic `imu/data` and will publish a new topic with the movement commands for turtlesim.

```
TeleopImu::TeleopImu()
{
pub = n.advertise<turtlesim::Velocity>("turtle1/command_velocity",1);
sub = n.subscribe<sensor_msgs::Imu>("imu/data", 10,
&TeleopImu::callBack, this);
}
```

The important part of the code is the `callBack` function. Inside this callback method, the IMU topic is received and processed to create a new `turtlesim/velocity` topic. As you might remember, this type of message will control the velocity of turtlesim.

```
void TeleopImu::callBack(const sensor_msgs::Imu::ConstPtr& imu)
{
...
tf::Matrix3x3(bq).getRPY(roll,pitch,yaw);
  vel.angular = roll;
    vel.linear = pitch;
pub.publish(vel);
}
```

The conversion of quaternions to Euler angles is done by the means of the `Matrix3x3` class. Then, we use the accessor method, `getRPY`, provided by this class. Using this, we will get the roll, pitch, and yaw from the matrix about the fixed x, y, and z axes.

After that, we only need to assign the value of pitch and roll to the linear and angular velocity variables, respectively.

Now, if you run everything at the same time, you will see the turtle moving according to the IMU movements as if it were a joystick.

Using a low-cost IMU – 10 degrees of freedom

In this section, we will learn to use a low-cost sensor with 10 **degrees of freedom** (**DoF**). This sensor, which is similar to Xsens MTi, has an accelerometer (x3), a magnetometer (x3), a barometer (x1), and a gyroscope (x3). It is controlled with a simple I2C interface, and in this example it will be connected to Arduino Nano (`http://arduino.cc/en/Main/ArduinoBoardNano`).

This sensor is also used for similar uses. Xsens costs approximately $2,000, which is very expensive for normal users. The sensor explained in this section has an approximate cost of $20. The low price of this sensor permits its usage in a lot of projects.

You can see this sensor in the following image. It is thin and has few components.

This board has the following sensors:

- **ADXL345**: This is a three-axis accelerometer with a high resolution (13-bit) measurement of up to ±16 g. This sensor is widely used in mobile device applications. It measures the static acceleration of gravity in tilt-sensing applications as well as dynamic acceleration resulting from motion or shock.
- **HMC5883L**: This sensor is designed for low-field magnetic sensing with a digital interface for devices such as low-cost compasses and magnetometers.
- **BMP085**: This is a high-precision barometric pressure sensor used in advanced mobile applications. It offers superior performance with an absolute accuracy up to to 0.03 hPa and has a very low power consumption of 3 μA.
- **L3G4200D**: This is a three-axis gyroscope with a very high resolution (16-bit) measurement up to 2,000 degrees per second (dps).This gyroscope measures how much the device is rotating around all three axes.

As we have said before, the board is controlled using the I2C protocol, and we will use Arduino to control it. In the following image, you can see the way to connect both the boards:

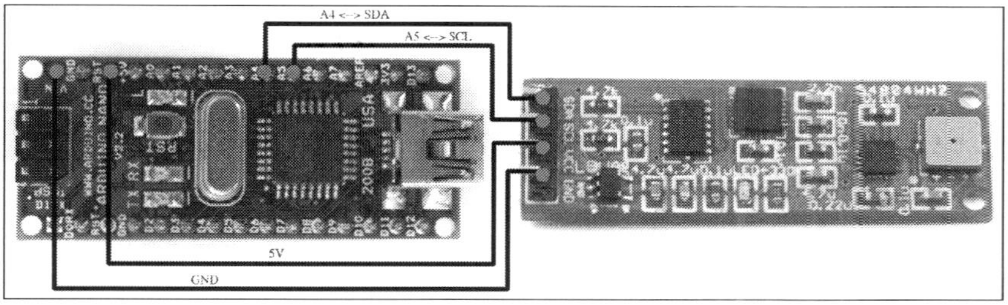

The only thing necessary is to connect the four wires to make it work. Connect the **GND** and **Vcc** from the sensor to **GND** and **5 V** in Arduino.

The **Serial Data Line** (**SDL**) must be connected to the analog pin 4 and the **Serial Clock** (**SCK**) must be connected to the analog pin 5. If you connect these pins wrongly, Arduino will not be able to communicate with the sensor.

Downloading the library for the accelerometer

Before using the sensor, it is necessary to download the right library for Arduino.

On the Internet, you can find a lot of libraries with different functionalities, but we will use the library that can be downloaded from `https://github.com/jenschr/Arduino-libraries`.

Once you have downloaded the library, uncompress it inside your `sketchbook` folder to load it.

You can find the libraries for the other sensors on the Internet. But to make your life easy, you can find all the necessary libraries to use the 10DOF sensor in `chapter4_tutorials/libraries`. Inside each library, you can also find examples about how to use the sensor.

Programming Arduino Nano and the 10DOF sensor

In this section, we are going to make a program in Arduino to take data from the accelerometers and publish it in ROS.

Open the Arduino IDE and create a new file with the name `example7.ino` and type in the following code:

```
#include <ros.h>
#include <std_msgs/Float32.h>

#include <Wire.h>
#include <ADXL345.h>

ADXL345 Accel;

ros::NodeHandle  nh;
std_msgs::Float32 velLinear_x;
std_msgs::Float32 velAngular_z;

ros::Publisher velLinear_x_pub("velLinear_x", &velLinear_x);
ros::Publisher velAngular_z_pub("velAngular_z", &velAngular_z);

void setup(){

  nh.initNode();
  nh.advertise(velLinear_x_pub);
  nh.advertise(velAngular_z_pub);

  Wire.begin();
  delay(100);
    Accel.set_bw(ADXL345_BW_12);

}

void loop(){

  double acc_data[3];
    Accel.get_Gxyz(acc_data);

  velLinear_x.data = acc_data[0];
  velAngular_z.data = acc_data[1];

  velLinear_x_pub.publish(&velLinear_x);
  velAngular_z_pub.publish(&velAngular_z);

  nh.spinOnce();
  delay(10);
}
```

Let's break the code and see the steps to take the data and publish it.

We are going to use two `Float 32` variables to publish the data, and it is necessary to include the following line in the program:

```
#include <std_msgs/Float32.h>
```

To be able to communicate with the sensor using the I2C Bus, we need to use the `Wire.h` library. This library is standard in Arduino.

```
#include <Wire.h>
```

To use the library downloaded before, we add the following `include` header:

```
#include <ADXL345.h>
```

We will use the `Accel` object to interact with the sensor.

```
ADXL345 Accel;
```

The data read from the sensor will be stored in the following variables: `velLinear_x` is used for linear velocity and the data is read from the accelerometer for the *x* axis, while `velAngular_z` is used for angular velocity and the data is read from the accelerometer for the *z* axis.

```
std_msgs::Float32 velLinear_x;
std_msgs::Float32 velAngular_z;
```

This program will publish two different topics: one for linear velocity and the other for angular velocity.

```
ros::Publisher velLinear_x_pub("velLinear_x", &velLinear_x);
ros::Publisher velAngular_z_pub("velAngular_z", &velAngular_z);
```

This is where the topic is created. Once you have executed these lines, you will see two topics in ROS with the names `velAngular_z` and `velLinear_x`.

```
nh.advertise(velLinear_x_pub);
nh.advertise(velAngular_z_pub);
```

The sensor and its bandwidth are initialized in this line as follows:

```
  Accel.set_bw(ADXL345_BW_12);
```

The data returned by `get_Gxyz` is stored in a three-component vector.

```
double acc_data[3];
  Accel.get_Gxyz(acc_data);
```

And finally, the data is published.

```
velLinear_x_pub.publish(&velLinear_x);
velAngular_z_pub.publish(&velAngular_z);
```

Creating a ROS node to use data from the 10DOF sensor

In this section, we are going to create a new program to use data from the sensor and generate moving commands to move the turtle in turtlesim. The goal is to use the 10DOF sensor board as input for turtlesim and moving the turtle using the accelerometers.

So, create a new file inside the directory `chapter4_tutorials/src` with the name `example8.cpp`, and type in the following code:

```
#include<ros/ros.h>
#include<turtlesim/Velocity.h>
#include<std_msgs/Float32.h>

class TeleopImu{
  public:
  TeleopImu();
  private:
    void velLinearCallBack(const std_msgs::Float32::ConstPtr& vx);
    void velAngularCallBack(const std_msgs::Float32::ConstPtr& wz);
    ros::NodeHandle n;
    ros::Publisher pub;
    ros::Subscriber velAngular_z_sub;
    ros::Subscriber velLinear_x_sub;
  turtlesim::Velocity vel;
};

TeleopImu::TeleopImu()
{
  pub = n.advertise<turtlesim::Velocity>("turtle1/command_
velocity",1);

  velLinear_x_sub = n.subscribe<std_msgs::Float32>("/velLinear_x", 1,
&TeleopImu::velLinearCallBack, this);
  velAngular_z_sub = n.subscribe<std_msgs::Float32>("/velAngular_z",
1, &TeleopImu::velAngularCallBack, this);
}
```

```
void TeleopImu::velAngularCallBack(const std_msgs::Float32::ConstPtr&
wz){
    vel.linear = -1 * wz->data;
    pub.publish(vel);
}

void TeleopImu::velLinearCallBack(const std_msgs::Float32::ConstPtr&
vx){
  vel.angular = vx->data;
    pub.publish(vel);
}

int main(int argc, char** argv)
{
  ros::init(argc, argv, "example8");
  TeleopImu teleop_turtle;
  ros::spin();
}
```

This code is similar to `example6.cpp` used with the Xsens IMU; the only difference being the topic to subscribe and the type of data. In this case, we will subscribe to two topics created by Arduino Nano. These topics will be used to control turtlesim.

It is necessary to subscribe to the topics, `/velLinear_x` and `velAngular_z`, and we will do this as shown in the following lines:

```
velLinear_x_sub = n.subscribe<std_msgs::Float32>("/velLinear_x", 1,
&TeleopImu::velLinearCallBack, this);
velAngular_z_sub = n.subscribe<std_msgs::Float32>("/velAngular_z", 1,
&TeleopImu::velAngularCallBack, this);
```

Every time the node receives a message, it takes the data from the message and creates a new `turtlesim/Velocity` message and publishes it.

```
void TeleopImu::velAngularCallBack(const std_msgs::Float32::ConstPtr&
wz){
  vel.linear = -1 * wz->data;
  pub.publish(vel);
}

void TeleopImu::velLinearCallBack(const std_msgs::Float32::ConstPtr&
vx){
vel.angular =  vx->data;
  pub.publish(vel);
}
```

To run the example, remember to compile the code and follow the steps outlined:

1. Start a new `roscore` command in a shell. Connect Arduino to the computer and launch the following command:

   ```
   $ rosrun rosserial_python serial_node.py
   ```

2. Now, start turtlesim by typing the following command:

   ```
   $ rosrun turtlesim turtlesim_node
   ```

And finally, type the following command to start the node:

```
$ rosrun chapter4_tutorials example8
```

If everything is OK, you should see the `TurtleSim` interface with the turtle moving. If you move the sensor, the turtle will move in a straight line or change its direction.

Summary

The use of sensors and actuators in robotics is very important since it is the only way to interact with the real world. In this chapter, you have learned how to use, configure, and investigate further on how some common sensors and actuators work, which are used by a number of people in the world of robotics. We are sure that if you wish to use another type of sensor, you will find information on the Internet about how to use it, without problems.

In our humble opinion, Arduino is a very interesting device because with it, you can add more devices or cheap sensors to your computer and use them within the ROS framework easily and transparently. Arduino has a large community, and you can find information on many sensors, which cover the spectrum of applications you can imagine.

Finally, we must mention that the range laser will be a very useful sensor in the upcoming chapters. The reason is that it is a mandatory device to implement the navigation stack, which relies on the range readings it provides at a high frequency with good precision. In the next chapter, you will see how to model your robot, visualize it in rviz, and use it by loading it on the Gazebo simulator, which is also integrated in ROS.

5
3D Modeling and Simulation

Programming directly on a real robot gives us good feedback and it is more impressive than simulations, but not everybody has possible access to real robots. For this reason, we have programs that simulate the physical world.

In this chapter we are going to learn how to:

- Create a 3D model of our robot
- Provide movements, physical limits, inertia, and other physical aspects to our robot
- Add simulated sensors to our 3D model
- Use the model on the simulator

A 3D model of our robot in ROS

The way ROS uses the 3D model of a robot or its parts, to simulate them or to simply help the developers in their daily work, is by means of the URDF files.

Unified Robot Description Format (**URDF**) is an XML format that describes a robot, its parts, its joints, dimensions, and so on. Every time you see a 3D robot on ROS, for example, the **PR2** (**Willow Garage**) or the **Robonaut** (**NASA**), a URDF file is associated with it. In the next sections we will learn how to create this file and the format to define different values.

Creating our first URDF file

The robot, which we are going to build in the following sections, is a mobile robot with four wheels and an arm with a gripper.

To start with, we create the base of the robots with four wheels. Create a new file in the `chapter5_tutorials/urdf` folder with the name `robot1.urdf` and put in the following code; this URDF code is based on XML, and the indentation is not mandatory but advisable, so use an editor that supports it or an adequate plugin or configuration (for example, an appropriate `.vimrc` file in Vim):

```
<?xml version="1.0"?>
  <robot name="Robot1">
    <link name="base_link">
      <visual>
        <geometry>
          <box size="0.2 .3 .1"/>
        </geometry>
        <origin rpy="0 0 0" xyz="0 0 0.05"/>
        <material name="white">
          <color rgba="1 1 1 1"/>
        </material>
      </visual>
    </link>

    <link name="wheel_1">
      <visual>
        <geometry>
          <cylinder length="0.05" radius="0.05"/>
        </geometry>
        <origin rpy="0 1.5 0" xyz="0.1 0.1 0"/>
          <material name="black">
            <color rgba="0 0 0 1"/>
          </material>
      </visual>
    </link>

    <link name="wheel_2">
      <visual>
        <geometry>
          <cylinder length="0.05" radius="0.05"/>
        </geometry>
        <origin rpy="0 1.5 0" xyz="-0.1 0.1 0"/>
        <material name="black"/>
      </visual>
    </link>
```

```
  <link name="wheel_3">
    <visual>
      <geometry>
        <cylinder length="0.05" radius="0.05"/>
      </geometry>
      <origin rpy="0 1.5 0" xyz="0.1 -0.1 0"/>
      <material name="black"/>
    </visual>
  </link>

  <link name="wheel_4">
    <visual>
      <geometry>
        <cylinder length="0.05" radius="0.05"/>
      </geometry>
      <origin rpy="0 1.5 0" xyz="-0.1 -0.1 0"/>
      <material name="black"/>
    </visual>
  </link>

  <joint name="base_to_wheel1" type="fixed">
    <parent link="base_link"/>
    <child link="wheel_1"/>
    <origin xyz="0 0 0"/>
  </joint>

  <joint name="base_to_wheel2" type="fixed">
    <parent link="base_link"/>
    <child link="wheel_2"/>
    <origin xyz="0 0 0"/>
  </joint>

  <joint name="base_to_wheel3" type="fixed">
    <parent link="base_link"/>
    <child link="wheel_3"/>
    <origin xyz="0 0 0"/>
  </joint>

  <joint name="base_to_wheel4" type="fixed">
    <parent link="base_link"/>
    <child link="wheel_4"/>
    <origin xyz="0 0 0"/>
  </joint>
</robot>
```

Explaining the file format

As you see in the code, there are two principal fields that describe the geometry of a robot: **links** and **joints**.

The first link has the name `base_link`; this name must be unique to the file:

```
<link name="base_link">
  <visual>
    <geometry>
      <box size="0.2 .3 .1"/>
    </geometry>
    <origin rpy="0 0 0" xyz="0 0 0.05"/>
    <material name="white">
      <color rgba="1 1 1 1"/>
    </material>
  </visual>
</link>
```

In order to define what we will see on the simulator, we use the visual field in the preceding code. Inside the code you can define the geometry (cylinder, box, sphere, or mesh), the material (color or texture), and the origin. We then have the code for the joint, shown as follows:

```
<joint name="base_to_wheel1" type="fixed">
      <parent link="base_link"/>
      <child link="wheel_1"/>
      <origin xyz="0 0 0"/>
</joint>
```

In the `joint` field, we define the name, which must be unique as well. Also, we define the type of joint (`fixed`, `revolute`, `continuous`, `floating`, or `planar`), the parent, and the child. In our case, `wheel_1` is a child of `base_link`. It is fixed, but as it is a wheel we can set it to `revolute`, for example.

To check whether the syntax is fine or whether we have errors, we can use the `check_urdf` command tool:

```
$ rosrun urdf_parser check_urdf robot1.urdf
robot name is: Robot1
---------- Successfully Parsed XML ---------------
root Link: base_link has 4 child(ren)
  child(1):  wheel_1
  child(2):  wheel_2
  child(3):  wheel_3
  child(4):  wheel_4
```

> This tool will be deprecated after ROS Fuerte, so if you are using a newer ROS distribution than the one used in this book (Fuerte), such as Groovy, you should use the `urdfdom` package, which will contain the `check_urdf` node/tool.

If you want to see it graphically, you can use the `urdf_to_graphiz` command tool:

```
$ rosrun urdf_parser urdf_to_graphiz "`rospack find chapter5_tutorials`/
urdf/robot1.urdf"
```

The following is what you will receive as output:

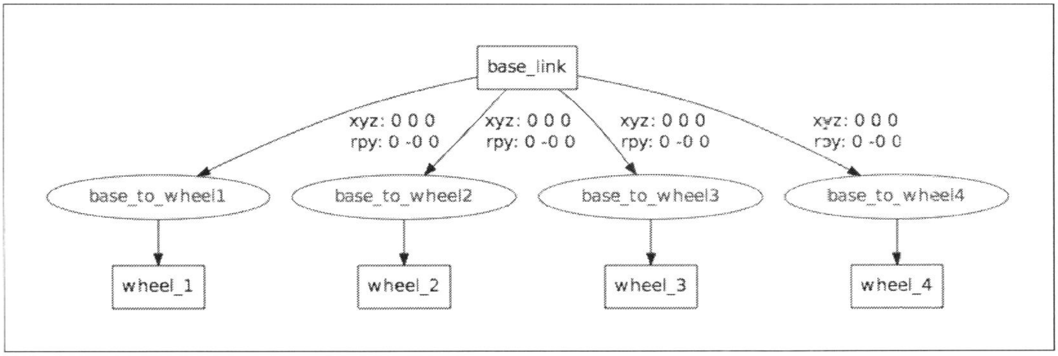

Watching the 3D model on rviz

Now that we have the model of our robot, we can use it on `rviz` to watch it in 3D and see the movements of the joints.

We will create the `display.launch` file in the `chapter5_tutorials/launch` folder, and put the following code in it:

```
<launch>
      <arg name="model" />
      <arg name="gui" default="False" />
      <param name="robot_description" textfile="$(arg model)" />
      <param name="use_gui" value="$(arg gui)"/>
      <node name="joint_state_publisher" pkg="joint_state_publisher"
type="joint_state_publisher" ></node>
      <node name="robot_state_publisher" pkg="robot_state_publisher"
type="state_publisher" />
      <node name="rviz" pkg="rviz" type="rviz" args="-d $(find urdf_
tutorial)/urdf.vcg" />
</launch>
```

We will launch it with the following command:

```
$ roslaunch chapter5_tutorials display.launch model:="`rospack find
chapter5_tutorials`/urdf/robot1.urdf"
```

If everything is fine, you will see the following window with the 3D model on it:

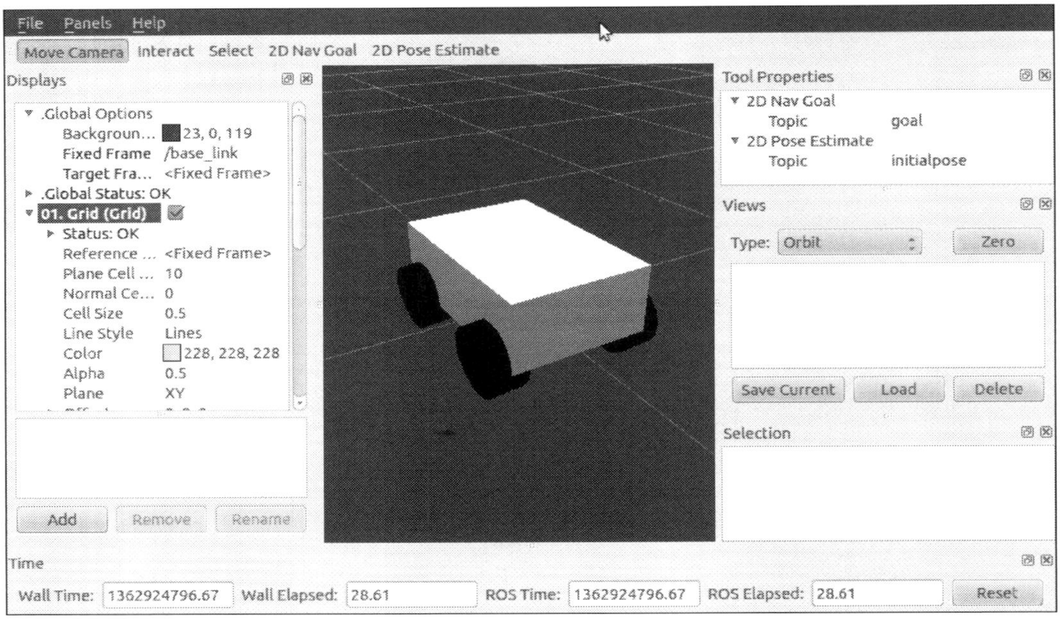

Let's finish the design by adding some parts: a base arm, an articulated arm, and a gripper. Try to finish the design yourself; you can find the final model in the `chapter5_tutorials/urdf/robot1.urdf` file. You can see the final design in the following screenshot as well:

> **Downloading the example code**
>
> You can download the example code files for all Packt books you have purchased from your account at `http://www.packtpub.com`. If you purchased this book elsewhere, you can visit `http://www.packtpub.com/support` and register to have the files e-mailed directly to you. You can also download these code files from `https://github.com/AaronMR/Learning_ROS_for_Robotics_Programming`.

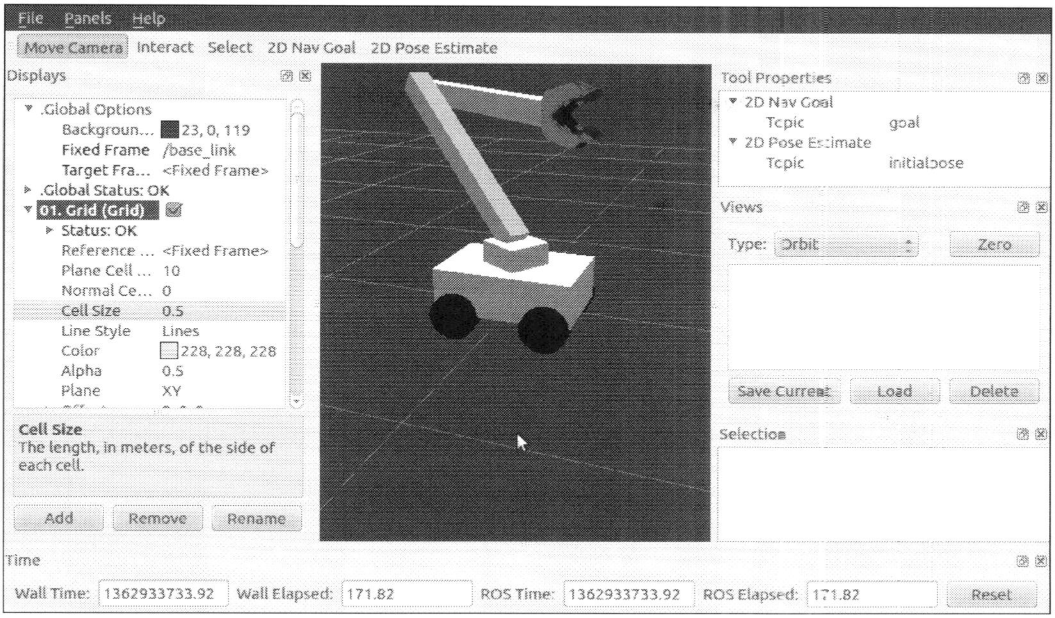

Loading meshes to our models

Sometimes we want to give more realistic elements to our model or make a more elaborate design, rather than using basic geometric objects/blocks. It is possible to load meshes generated by us or to use meshes of other models. For our model, we used the PR2's gripper. In the following code, you can see an example of how to use it:

```
<link name="left_gripper">
<visual>
       <origin rpy="0 0 0" xyz="0 0 0"/>
           <geometry>
             <mesh filename="package://pr2_description/meshes/gripper_
v0/l_finger.dae"/>
          </geometry>
     </visual>
</link>
```

This looks like the sample link that we used before, but in the geometry section we added the mesh we are going to use. You can see the result in the following screenshot:

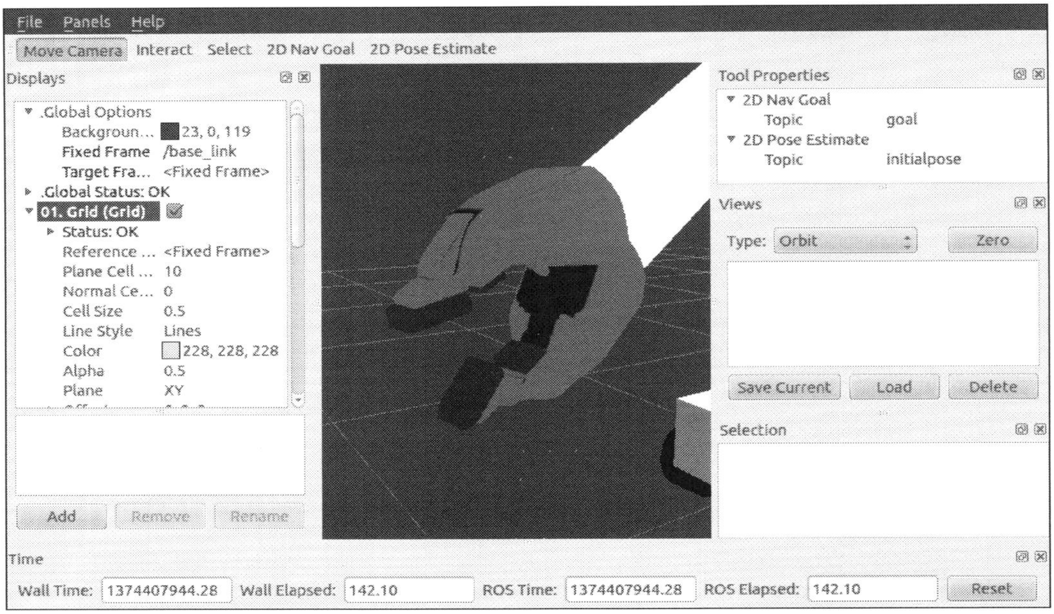

Making our robot model movable

To convert the model into a robot that can actually move, the only thing you have to do is take care of the type of the joints it uses. If you check the URDF model file, you will see the different types of joints used in this model.

The most used type of joint is the revolute joint. For example, the one used on `arm_1_to_arm_base`, is shown in the following code:

```
<joint name="arm_1_to_arm_base" type="revolute">
              <parent link="arm_base"/>
              <child link="arm_1"/>
              <axis xyz="1 0 0"/>
              <origin xyz="0 0 0.15"/>
<limit effort ="1000.0" lower="-1.0" upper="1.0" velocity="0.5"/>
</joint>
```

This means that they rotate in the same way that the continuous joints do, but they have strict limits. The limits are fixed using the `<limit effort ="1000.0" lower="-1.0" upper="1.0" velocity="0.5"/>` line, and you can select the axis to move with `axis xyz="1 0 0"`.

A good way of testing whether or not the axis and limits of the joints are fine is by running `rviz` with the `Join_State_Publisher` GUI:

```
$ roslaunch chapter5_tutorials display.launch model:="`rospack find chapter5_tutorials`/urdf/robot1.urdf" gui:=true
```

You will see the `rviz` interface with another window with some sliders, each one controlling one joint:

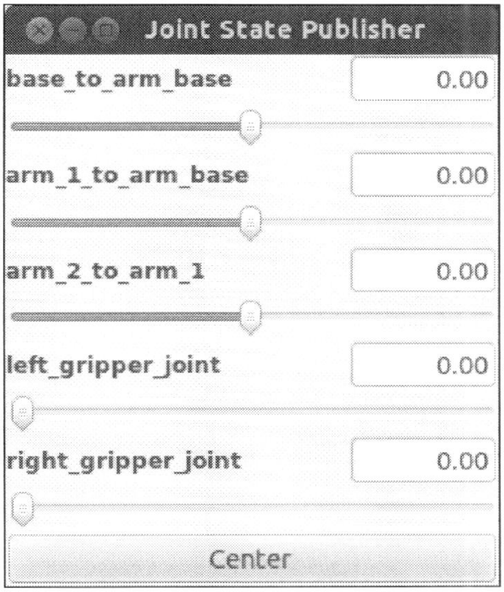

Physical and collision properties

If you want to simulate the robot on **Gazebo** or any other simulation software, it is necessary to add physical and collision properties. This means that we need to set the dimension of the geometry to calculate the possible collisions, for example, the weight that will give us the inertia, and so on.

It is necessary that all links on the model file have these parameters; if not, the robot could not be simulated.

For the mesh models, it is easier to calculate collisions by using simplified geometry than the actual mesh. Calculating the collision between two meshes is more computationally complex than it is to calculate a simple geometry.

In the following code, you will see the new parameters added on the link with the name `wheel_1`:

```
<link name="wheel_1">
  ...
  <collision>
    <geometry>
      <cylinder length="0.05" radius="0.05"/>
    </geometry>
  </collision>
  <inertial>
    <mass value="10"/>
    <inertia ixx="1.0" ixy="0.0" ixz="0.0" iyy="1.0" iyz="0.0"
izz="1.0"/>
  </inertial>
</link>
```

It is the same for the other links. Remember to put `collision` and `inertial` elements in all the links, because if you do not, Gazebo will not take the model.

You can find a complete file with all the parameters at `chapter5_tutorials/urdf/robot1_physics.urdf`.

Xacro – a better way to write our robot models

Notice the size of the `robot1_physics.urdf` file. It has 314 lines of code to define our robot. Imagine if you start to add cameras, legs, and other geometries, the file will start to increase and the maintenance of the code will start to become more complicated.

Xacro helps to reduce the overall size of the URDF file and makes it easier to read and maintain. It also allows us to create modules and reutilize them to create repeated structures such as several arms or legs.

To start using `xacro`, we need to specify a namespace so that the file is parsed properly. For example, these are the first two lines of a valid `xacro` file:

```
<?xml version="1.0"?>
<robot xmlns:xacro="http://www.ros.org/wiki/xacro" name="robot1_
xacro">
```

In it, we define the name of the model, which in this case is `robot1_xacro`. Remember that the file extension will be `.xacro` instead of `.urdf`.

Using constants

We can use xacro to declare constant values; hence we can avoid putting the same value in a lot of lines. Without the use of xacro, if we had to change a value, it would be almost impossible to maintain the changes.

For example, the four wheels have the same values for length and radius. If we want to change the value, we need to change it in each line; but if we use the next lines, we can change all the values easily:

```
<xacro:property name="length_wheel" value="0.05" />
<xacro:property name="radius_wheel" value="0.05" />
```

And now, to use these variables you only have to change the old value with `${name_of_variable}`:

```
<cylinder length="${length_wheel}" radius="${radius_wheel}"/>
```

Using math

You can build up arbitrarily complex expressions in the `${}` construct using the four basic operations (+, -, *, /), the unary minus, and parenthesis. Exponentiation and modulus are, however, not supported:

```
<cylinder radius="${wheeldiam/2}" length=".1"/>
<origin xyz="${reflect*(width+.02)} 0 .25" />
```

By using mathematics, we can resize the model by only changing a value. To do this, we need a parameterized design.

Using macros

Macros are the most useful component of the xacro package. To reduce the file size even more, we are going to use the following macro for `inertial`:

```
<xacro:macro name="default_inertial" params="mass">
              <inertial>
                      <mass value="${mass}" />
                      <inertia ixx="1.0" ixy="0.0" ixz="0.0"
                               iyy="1.0" iyz="0.0"
                               izz="1.0" />
              </inertial>
      </xacro:macro>
<xacro:default_inertial mass="100"/>
```

If we compare the `robot1.urdf` file with `robot1.xacro`, we will have eliminated 30 duplicate lines without effort. It is possible to reduce it further using more macros and variables.

To use the `xacro` file with `rviz` and Gazebo, you need to convert it to `.urdf`. To do this, we execute the following command inside the `chapter5_tutorials/urdf` folder:

```
$ rosrun xacro xacro.py robot1.xacro > robot1_processed.urdf
```

You can also execute the following command everywhere and it should give the same result as the other command:

```
$ rosrun xacro xacro.py "`rospack find chapter5_tutorials`/urdf/robot1.
xacro" > "`rospack find chapter5_tutorials`/urdf/robot1_processed.urdf"
```

So, in order to make the commands easier to write, we recommend you to continue working on the same folder.

Moving the robot with code

Ok, we have the 3D model of our robot and we can see it on `rviz`, but how can we move the robot using a node?

Create a new file in the `chapter5_tutorials/src` folder with the name `state_publisher.cpp` and copy the following code:

```
#include <string>
#include <ros/ros.h>
#include <sensor_msgs/JointState.h>
#include <tf/transform_broadcaster.h>

int main(int argc, char** argv) {
  ros::init(argc, argv, "state_publisher");
  ros::NodeHandle n;
  ros::Publisher joint_pub = n.advertise<sensor_
msgs::JointState>("joint_states", 1);
  tf::TransformBroadcaster broadcaster;
  ros::Rate loop_rate(30);

  const double degree = M_PI/180;

  // robot state
  double inc= 0.005, base_arm_inc= 0.005, arm1_armbase_inc= 0.005,
arm2_arm1_inc= 0.005, gripper_inc= 0.005, tip_inc= 0.005;
```

```
  double angle= 0 ,base_arm = 0, arm1_armbase = 0, arm2_arm1 = 0,
gripper = 0, tip = 0;
  // message declarations
  geometry_msgs::TransformStamped odom_trans;
  sensor_msgs::JointState joint_state;
  odom_trans.header.frame_id = "odom";
  odom_trans.child_frame_id = "base_link";

  while (ros::ok()) {
      //update joint_state
      joint_state.header.stamp = ros::Time::now();
      joint_state.name.resize(7);
      joint_state.position.resize(7);
      joint_state.name[0] ="base_to_arm_base";
      joint_state.position[0] = base_arm;
      joint_state.name[1] ="arm_1_to_arm_base";
      joint_state.position[1] = arm1_armbase;
      joint_state.name[2] ="arm_2_to_arm_1";
      joint_state.position[2] = arm2_arm1;
      joint_state.name[3] ="left_gripper_joint";
      joint_state.position[3] = gripper;
      joint_state.name[4] ="left_tip_joint";
      joint_state.position[4] = tip;
      joint_state.name[5] ="right_gripper_joint";
      joint_state.position[5] = gripper;
      joint_state.name[6] ="right_tip_joint";
      joint_state.position[6] = tip;

      // update transform
      // (moving in a circle with radius 1)
      odom_trans.header.stamp = ros::Time::now();
      odom_trans.transform.translation.x = cos(angle);
      odom_trans.transform.translation.y = sin(angle);
      odom_trans.transform.translation.z = 0.0;
      odom_trans.transform.rotation = tf::createQuaternionMsgFromYaw(
angle);

        //send the joint state and transform
      joint_pub.publish(joint_state);
      broadcaster.sendTransform(odom_trans);

        // Create new robot state
      arm2_arm1 += arm2_arm1_inc;
```

```
        if (arm2_arm1<-1.5 || arm2_arm1>1.5) arm2_arm1_inc *= -1;
          arm1_armbase += arm1_armbase_inc;
  if (arm1_armbase>1.2 || arm1_armbase<-1.0) arm1_armbase_inc *= -1;
        base_arm += base_arm_inc;
        if (base_arm>1. || base_arm<-1.0) base_arm_inc *= -1;
        gripper += gripper_inc;
        if (gripper<0 || gripper>1) gripper_inc *= -1;
         angle += degree/4;

        // This will adjust as needed per iteration
        loop_rate.sleep();
    }
    return 0;
  }
```

We are going to see what we can do to the code to get these movements.

First, we create a new frame called `odom` and all the transforms will be referred to this new frame. As you will remember, all the links are children of `base_link` and all the frames will be linked to the `odom` frame:

```
...
geometry_msgs::TransformStamped odom_trans;
odom_trans.header.frame_id = "odom";
odom_trans.child_frame_id = "base_link";
...
```

Now, we are going to create a new topic to control all the joints of the model. `Joint_state` is a message that holds data to describe the state of a set of torque-controlled joints. As our model has seven joints, we create a message with seven elements:

```
sensor_msgs::JointState joint_state;

joint_state.header.stamp = ros::Time::now();
joint_state.name.resize(7);
joint_state.position.resize(7);
joint_state.name[0] ="base_to_arm_base";
joint_state.position[0] = base_arm;
...
```

In our example, the robot will move in circles. We calculate the coordinates and the movement on the next portion of our code:

```
odom_trans.header.stamp = ros::Time::now();
odom_trans.transform.translation.x = cos(angle)*1;
odom_trans.transform.translation.y = sin(angle)*1;
odom_trans.transform.translation.z = 0.0;
odom_trans.transform.rotation = tf::createQuaternionMsgFromYaw(angle);
```

Finally, we publish the new state of our robot:

```
joint_pub.publish(joint_state);
broadcaster.sendTransform(odom_trans);
```

We are also going to create a launch file to launch the node, the model, and all the necessary elements. Create a new file with the name `display_xacro.launch` (content given as follows) and put it in the `chapter5_tutorials/launch` folder:

```
<launch>
      <arg name="model" />
      <arg name="gui" default="False" />
      <param name="robot_description" command="$(find xacro)/xacro.py
$(arg model)" />
      <param name="use_gui" value="$(arg gui)"/>
      <node name="state_publisher" pkg="chapter5_tutorials"
type="state_publisher" />
      <node name="robot_state_publisher" pkg="robot_state_publisher"
type="state_publisher" />
      <node name="rviz" pkg="rviz" type="rviz" args="-d $(find
chapter5_tutorial)/urdf.vcg" />
</launch>
```

Using the following command, we will start our new node with the complete model. We will see the 3D model on `rviz` moving all the articulations:

```
$ roslaunch chapter5_tutorials state_xacro.launch model:="`rospack find
chapter5_tutorials`/urdf/robot1.xacro"
```

In the following screenshot, you can see a mix of four screens captured to show you the movements that we obtained with the node. If you see it fine, you will see the trajectory through a circle and the arms moving:

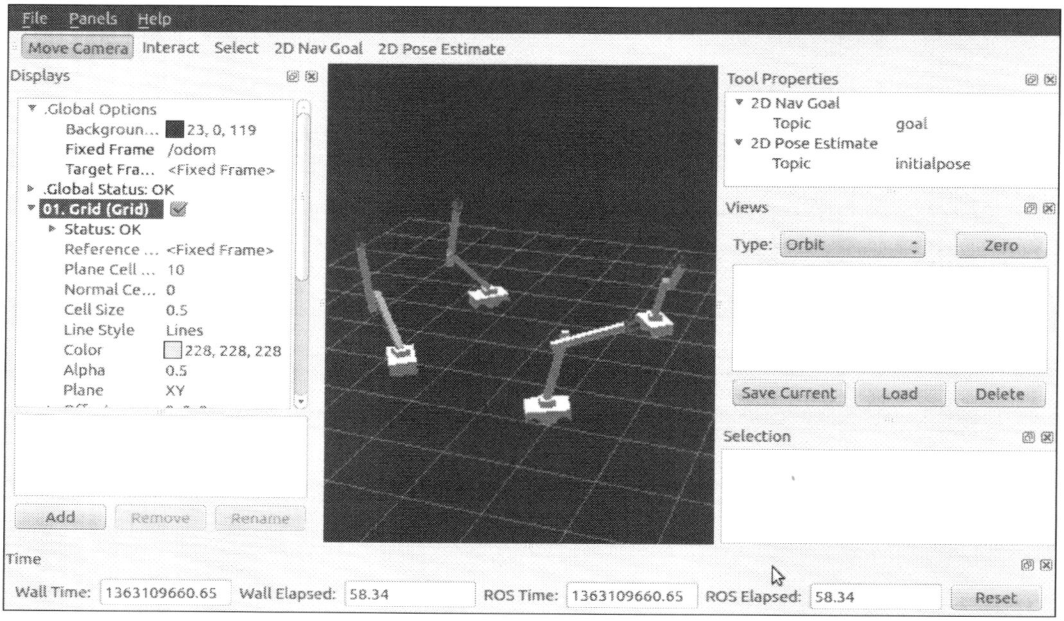

3D modeling with SketchUp

It is possible to generate the model using 3D programs such as `SketchUp`. In this section we will show you how to make a simple model, export it, generate an `urdf`, and watch the model on `rviz`. Notice that SketchUp works on Windows and Mac, and that this model was developed using Mac and not Linux.

First, you need to have **SketchUp** installed on your computer. When you have it, make a model similar to the following:

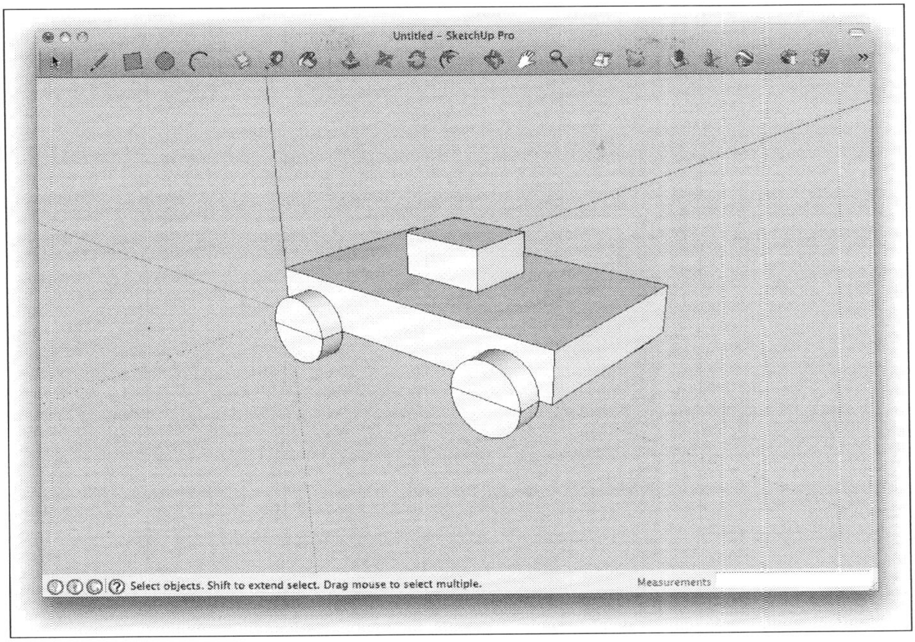

The model was exported only to one file, so the wheels and chassis are the same object. If you want to make a robot model with mobile parts, you must export each part in a separate file.

To export the model, navigate to **Export | 3D Model | Save As COLLADA File** (`*.dae`).

We named the file `bot.dae` and we saved it in the `chapter5_tutorials/urdf` folder.

Now, to use the 3D model, we are going to create a new file in the `chapter5_tutorials/urdf` folder with the name `dae.urdf` and type in the following code:

```
<?xml version="1.0"?>
<robot name="robot1">
  <link name="base_link">
  <visual>
    <geometry>
      <mesh scale="0.025 0.025 0.025" filename="package://chapter5_
tutorials/urdf/bot.dae"/>
    </geometry>
    <origin xyz="0 0 0.226"/>
  </visual>
  </link>
</robot>
```

Test the model with the following command:

```
$ roslaunch chapter5_tutorials display.launch model:="`rospack find
chapter5_tutorials`/urdf/dae.urdf"
```

You will see the following output:

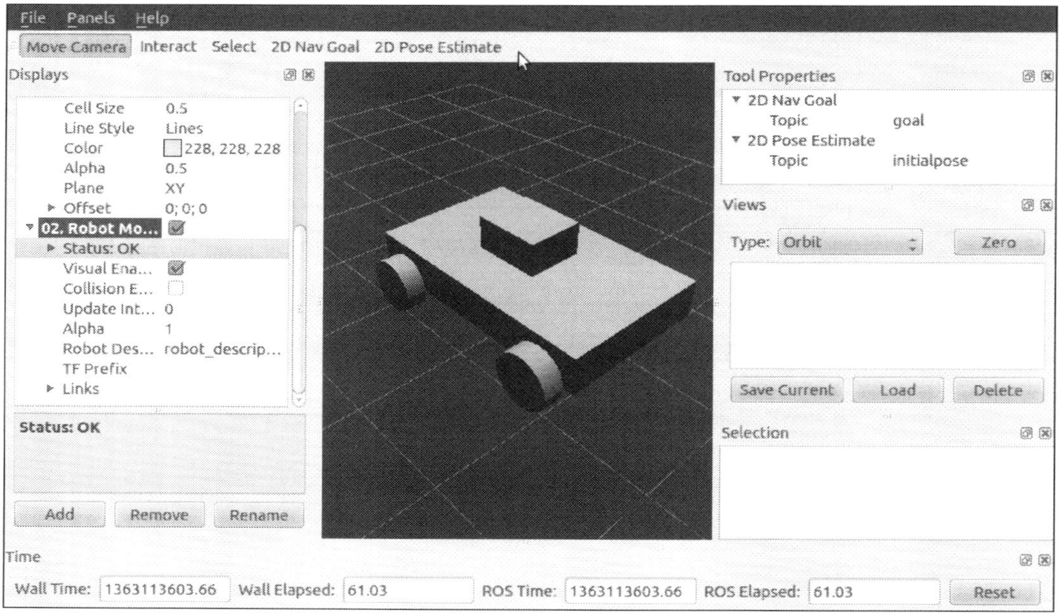

Simulation in ROS

In order to make simulations with our robots on ROS, we are going to use Gazebo.

Gazebo (`http://gazebosim.org/`) is a multirobot simulator for outdoor environments. It is capable of simulating a population of robots, sensors, and objects in a three-dimensional world. It generates both realistic sensor feedback and physically plausible interactions between objects.

In this section you will learn how to use the model created before, how to include a laser sensor, and how to move it as a real robot.

Using our URDF 3D model in Gazebo

We are going to use the model that we designed in the last section, but without the arm, to make it simple.

It is necessary to complete the URDF model because in order to use it in Gazebo we need to declare more elements. We will also use the `.xacro` file; although this may be more complex, it is more powerful for development of the code. You have a file with all the modifications at `chapter5_tutorials/urdf/robot1_base_01.xacro`:

```
<link name="base_link">
  <visual>
    <geometry>
      <box size="0.2 .3 .1"/>
    </geometry>
    <origin rpy="0 0 1.54" xyz="0 0 0.05"/>
    <material name="white">
      <color rgba="1 1 1 1"/>
    </material>
  </visual>
  <collision>
    <geometry>
      <box size="0.2 .3 0.1"/>
    </geometry>
  </collision>
  <xacro:default_inertial mass="10"/>
</link>
```

This is the new code for the chassis of the robot `base_link`. Notice that the `collision` and `inertial` sections are necessary to run the model on Gazebo in order to calculate the physics of the robot.

To launch everything, we are going to create a new `.launch` file. Create a new file with the name `gazebo.launch` in the `chapter5_tutorials/launch` folder, and put in the following code:

```
<?xml version="1.0"?>
<launch>
<param name="/use_sim_time" value="true" />
    <include file="$(find gazebo_worlds)/launch/empty_world.launch"/>
    <arg name="model" />
    <param name="robot_description" command="$(find xacro)/xacro.py
$(arg model)" />
    <node name="joint_state_publisher" pkg="joint_state_publisher"
type="joint_state_publisher" ></node>
```

```
    <node pkg="robot_state_publisher" type="state_publisher"
name="robot_state_publisher" output="screen" >
  <param name="publish_frequency" type="double" value="50.0" />
   </node>
    <node name="spawn_robot" pkg="gazebo" type="spawn_model"
args="-urdf -param robot_description -z 0.1 -model robot_model"
respawn="false" output="screen" />
</launch>
```

To launch the file, use the following command:

```
$ roslaunch chapter5_tutorials gazebo.launch  model:="`rospack find
chapter5_tutorials`/urdf/robot1_base_01.xacro"
```

You will now see the robot in Gazebo. Congratulations! This is your first step in the virtual world:

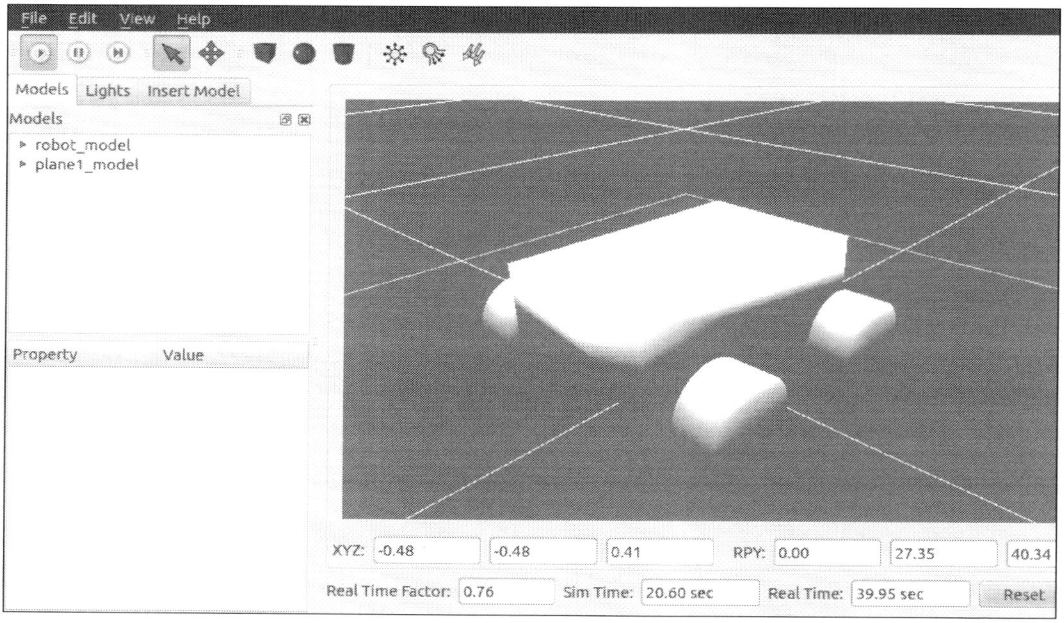

As you can see, the model has no texture. In `rviz`, you observed the textures that were declared in the URDF file. But in Gazebo, you cannot see them.

To add textures visible in Gazebo, use the following code on your model file. Copy the `chapter5_tutorials/urdf/robot1_base_01.xacro` file and save it with the name `robot1_base_02.xacro` and add the following code inside:

```
<gazebo reference="base_link">
      <material>Erratic/BlueBrushedAluminum</material>
 </gazebo>

 <gazebo reference="wheel_1">
      <material>Erratic/Black</material>
 </gazebo>

 <gazebo reference="wheel_2">
      <material>Erratic/Black</material>
 </gazebo>

 <gazebo reference="wheel_3">
      <material>Erratic/Black</material>
 </gazebo>

 <gazebo reference="wheel_4">
      <material>Erratic/Black</material>
 </gazebo>
```

Launch the new file and you will see the same robot, but with the added textures:

```
$ roslaunch chapter5_tutorials gazebo.launch  model:="`rospack find chapter5_tutorials`/urdf/robot1_base_02.xacro"
```

You will see the following output:

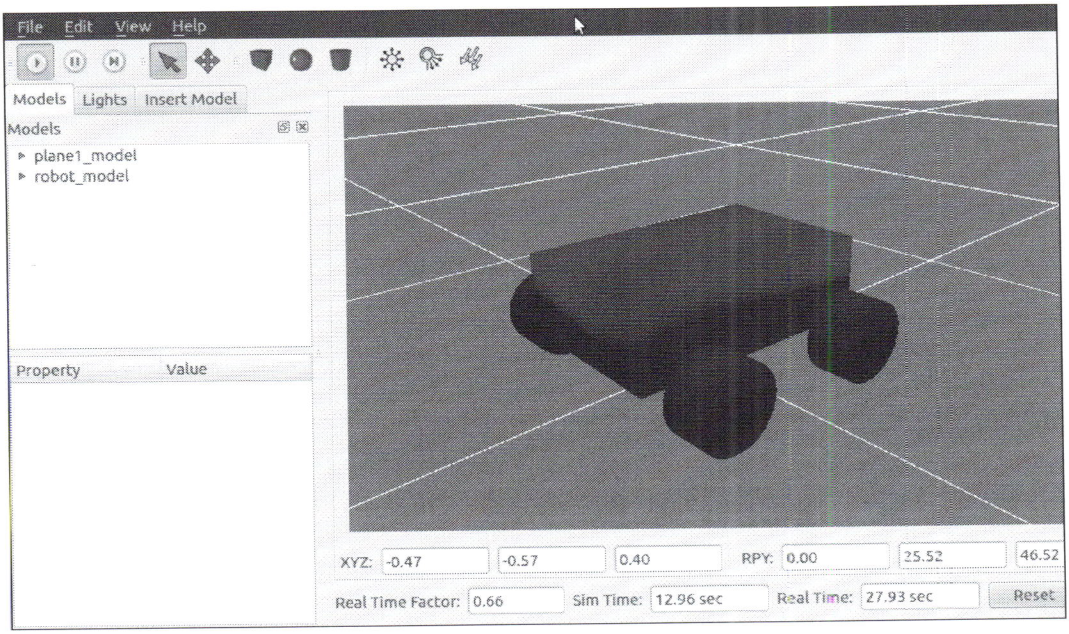

Adding sensors to Gazebo

In Gazebo, you can simulate the physics of the robot and its movement, and you can also simulate sensors.

Normally, when you want to add a new sensor you need to implement the behavior. Fortunately, some sensors are already developed for Gazebo and ROS.

In this section we are going to add a laser sensor to our model. This sensor will be a new part on the robot. You need to select where to put it. In Gazebo, you will see a new 3D object that looks like a Hokuyo laser. We talked about this sensor in the past chapters.

We are going to take the laser from the `erratic_description` package. This is the magic of ROS, you can re-use code from other packages for your development.

Only, it is necessary to add the next lines on our `.xacro` file:

```
<include filename="$(find erratic_description)/urdf/erratic_hokuyo_
laser.xacro" />
<!-- BASE LASER ATTACHMENT -->
<erratic_hokuyo_laser parent="base_link">
  <origin xyz="0.18 0 0.11" rpy="0 0 0" />
</erratic_hokuyo_laser>
```

Launch the new model with the following command:

```
$ roslaunch chapter5_tutorials gazebo.launch model:="`rospack find
chapter5_tutorials`/urdf/robot1_base_03.xacro"
```

You will see the robot with the laser module attached to it, as shown in the following screenshot:

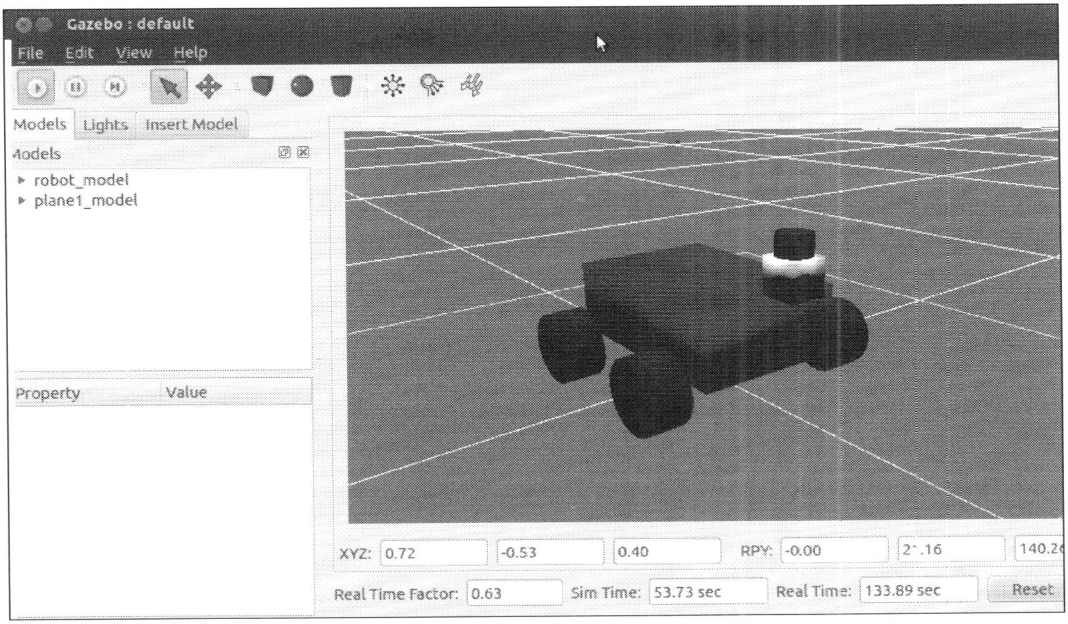

Notice that this laser is generating "real" data as a real laser. You can see the data generated using the `rostopic echo` command:

```
$ rostopic echo /base_scan/scan
```

Loading and using a map in Gazebo

In Gazebo, you can use virtual worlds as offices, mountains, and so on.

In this section we are going to use a map of the office of Willow Garage that is installed by default with the ROS installation.

This 3D model is in the `gazebo_worlds` package. If you do not have the package, install it before you continue.

To check the model, you will only have to start the `.launch` file using the following command:

```
$ roslaunch gazebo_worlds wg_collada_world.launch
```

You will see the 3D office in Gazebo. The office only has walls. You can add tables, chairs, and much more, if you want. Please note that Gazebo requires a good machine with a relatively recent GPU. You can check whether your graphics are supported at the Gazebo homepage. Also, note that sometimes this software crashes, but great effort is being taken by the community to make it more stable. Usually, it is enough to run it again (probably several times) if it crashes. If the problem persists, our advice is to try with a newer version, which will be installed by default with more recent distributions of ROS.

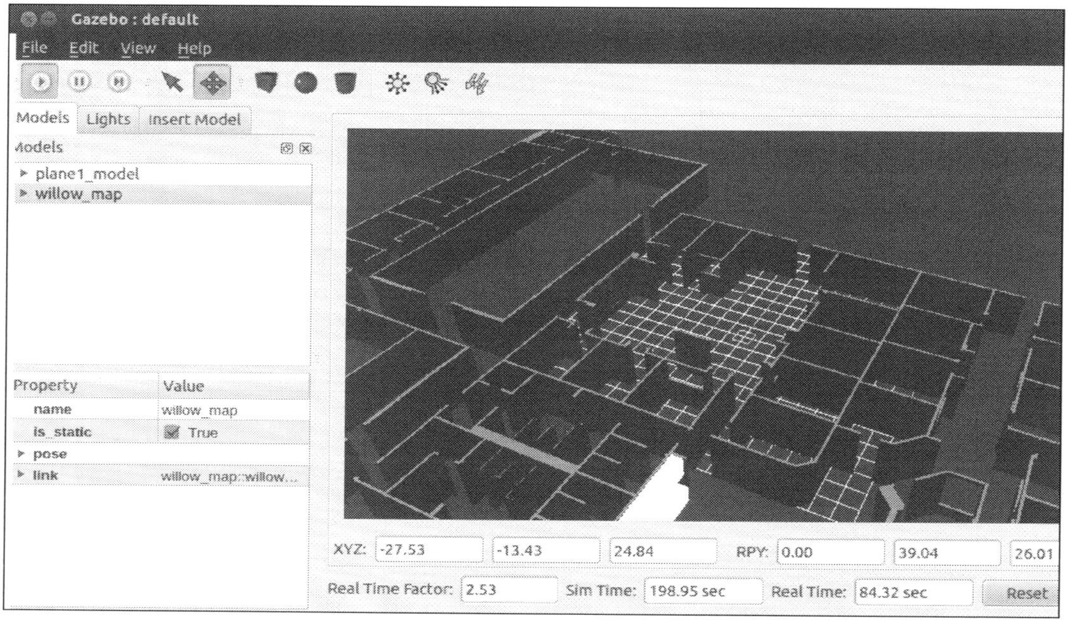

What we are going to do now is create a new `.launch` file to load the map and the robot together. To do that, create a new file in the `chapter5_tutorials/launch` folder with the name `gazebo_map_robot.launch` and add the following code:

```
<?xml version="1.0"?>
<launch>
  <!-- this launch file corresponds to robot model in ros-pkg/robot_
descriptions/pr2/erratic_defs/robots for full erratic -->

  <param name="/use_sim_time" value="true" />

  <!-- start up wg world -->
  <include file="$(find gazebo_worlds)/launch/wg_collada_world.
launch"/>
```

```
  <arg name="model" />
  <param name="robot_description" command="$(find xacro)/xacro.py
$(arg model)" />

  <node name="joint_state_publisher" pkg="joint_state_publisher"
type="joint_state_publisher" ></node>
  <!-- start robot state publisher -->
  <node pkg="robot_state_publisher" type="state_publisher"
name="robot_state_publisher" output="screen" >
    <param name="publish_frequency" type="double" value="50.0" />
  </node>

  <node name="spawn_robot" pkg="gazebo" type="spawn_model" args="-urdf
-param robot_description -z 0.1 -model robot_model" respawn="false"
output="screen" />
</launch>
```

Now launch the file of the model with the laser:

```
$ roslaunch chapter5_tutorials gazebo_map_robot.launch  model:="`rospack
find chapter5_tutorials`/urdf/robot1_base_03.xacro"
```

You will see the robot and the map on the Gazebo GUI. The next step is to command the robot to move and receive the simulated readings of its sensors as it evolves around the virtual world loaded in the simulator.

Moving the robot in Gazebo

A differential wheeled robot is a mobile robot whose movement is based on two separately driven wheels placed on either side of the robot body. It can thus change its direction by varying the relative rate of rotation of its wheels and does not require an additional steering motion.

As we said before, in Gazebo you need to program the behaviors of the robot, joints, sensors, and so on. As it happened for the laser, Gazebo already has a differential drive implemented and we can use it to move our robot.

Before we continue, the `erratic_robot` stack is necessary for this section. The example will use the differential erratic robot drive from `erratic_robot` to move the 3D model in `gazebo`. To install the stack, use the following command:

```
$ sudo apt-get install ros-fuerte-erratic-robot
```

The differential drive is usually meant for robots with only two wheels, but our robot has four wheels. So, we have a problem with the movement, since it will not be correct.

Anyway, we are going to use it for our robot, since the modification is simple. In fact, you can take the code of the differential drive and modify it to add four wheels.

To use this controller, you only have to add the following code to the model file:

```
<gazebo>
<controller:diffdrive_plugin name="differential_drive_controller"
plugin="libdiffdrive_plugin.so">
      <alwaysOn>true</alwaysOn>
      <update>100</update>
      <updateRate>100.0</updateRate>
      <leftJoint>base_to_wheel4</leftJoint>
      <rightJoint>base_to_wheel1</rightJoint>
      <wheelSeparation>0.33</wheelSeparation>
      <wheelDiameter>0.1</wheelDiameter>
      <torque>5</torque>
      <interface:position name="position_iface_0"/>
      <topicName>cmd_vel</topicName>
</controller:diffdrive_plugin>
</gazebo>
```

The parameters that you can see in the code are simply the configuration set up to make the controller work with our four-wheeled robot.

For example, we selected the `base_to_wheel4` and `base_to_wheel1` joints as the principal wheels to move the robot. Remember that we have four wheels, but the controller only needs two of them.

Another interesting parameter is `topicName`. We need to publish commands with this name in order to control the robot.

It is necessary to add another link at the beginning of the model. Before, the first link was `base_link`, now the first link is `base_footprint`.

The controller will make the transformation between this link and it will create another link called `odom`. The `odom` link will be used with the navigation stack in future chapters:

```
<link name="base_footprint">
  <visual>
      <geometry>
            <box size="0.001 0.001 0.001"/>
      </geometry>
```

```
        <origin rpy="0 0 0" xyz="0 0 0"/>
  </visual>
  <xacro:default_inertial mass="0.0001"/>
</link>

<gazebo reference="base_footprint">
  <material>Gazebo/Green</material>
    <turnGravityOff>false</turnGravityOff>
</gazebo>

<joint name="base_footprint_joint" type="fixed">
   <origin xyz="0 0 0" />
   <parent link="base_footprint" />
   <child link="base_link" />
</joint>
```

All these changes are in the `chapter5_tutorials/urfd/robot1_base_04.xacro` file.

Now, to launch the model with the controller and the map, we use the following command:

```
$ roslaunch chapter5_tutorials gazebo_map_robot.launch model:="`rospack find chapter5_tutorials`/urdf/robot1_base_04.xacro"
```

You will see the map with the robot on the Gazebo screen:

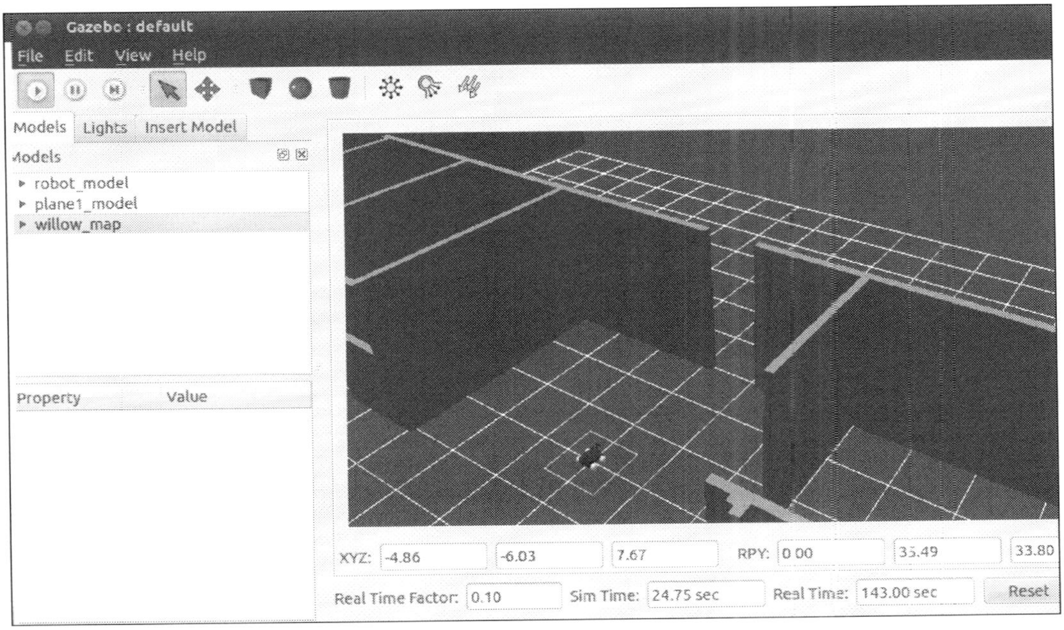

Now we are going to move the robot using a node. This node is in the `erratic_teleop` package.

Run the following command:

```
$ rosrun erratic_teleop erratic_keyboard_teleop
```

You will see a new shell with some instructions and WASD keys to move the robot.

If you start `rviz` now, you will see the 3D model and the laser data on the screen. If you move the robot, you will see the robot moving on `rviz`.

Summary

For people learning robotics, the ability to have access to real robots is fun and useful, but not everyone has access to a real robot. This is why simulators exist.

In this chapter you have learned how to create a 3D model of your own robot. This includes a detailed explanation that guides you in the tasks of adding textures and creating joints, and also describes how to move the robot with a node.

Then we have introduced Gazebo, a simulator where you can load the 3D model of your robot and simulate it moving and sensing a virtual world. This simulator is widely used by the ROS community and it already supports many real robots in simulation.

Indeed, we have seen how to re-use parts of other robots to design ours in a nutshell. In particular, we have included a gripper and added sensors, such as a laser range finder.

Hence, it is not mandatory to create a robot from scratch to start using the simulator. The community has developed a lot of robots and you can download the code, execute them in ROS and Gazebo, and modify them if it turns out to be necessary.

You can find a list of the robots supported on ROS on `http://www.ros.org/wiki/Robots`.

In the next chapter, we will learn about computer vision and packages, such as SLAM and visual recognition, to perform various tasks with cameras.

6
Computer Vision

In ROS, the support for computer vision is provided by means of camera drivers, the integration of OpenCV libraries, tools to set the frame transform (`tf`) of the camera optical frame with respect to our robot, and a good number of third-party tools, which comprise algorithms for visual odometry, augmented reality, object detection, and perception, among others.

The first capability that ROS offers us when working with vision is the ability to manage FireWire (IEEE1394a or IEEE1394b) cameras. Indeed, a package in the main ROS framework contains the drivers for such cameras. In the case of USB or gigabit Ethernet cameras, we must install third-party drivers or use our own. In this chapter, we will list USB drivers that come as ROS packages, and we will also provide a driver source code implemented with the OpenCV video capture API. USB cameras are pretty inexpensive and easy to find on the market, so they follow the goal of this book: learn by practice.

This chapter follows the ensuing outline:

- First, we will see how to connect and run cameras in ROS, starting with FireWire cameras, which have the best support in ROS. But then we will also cover the different options that we have to work with USB cameras, which are easier to find and are cheaper; indeed, we will also explain how to implement a camera driver with OpenCV for USB cameras.
- The USB camera driver will use most of the APIs of OpenCV. We will have a look at how to connect the frame captured with ROS images, by using the **cv_bridge** package, which is explained in detail. Similarly, we will introduce the ImageTransport API that allows publishing an image in several formats with or without compression.

> OpenCV is probably the most used open source computer vision library out there. ROS comes with the binaries of one of the latest stable versions (or just depends on the OpenCV library Debian package installed in your system in recent ROS distributions), so we can easily use it in our nodes. In order to subscribe and publish images, the OpenCV image format can be transformed to the ROS image package by means of the `cv_bridge` and `image_transport` packages, whose APIs will be explained later in this chapter.

Also, we will see how to configure the camera parameters, a feature that is specially supported in FireWire cameras.

- Having OpenCV inside ROS, we have the ability to implement a wide range of computer vision and machine learning algorithms, or even to run some algorithms or examples already present in this library. Here, we will not discuss the OpenCV API as it is outside the scope of this book. Alternatively, we advise the reader to check the online documentation (`http://docs.opencv.org`) or any book about OpenCV and computer vision itself. Just remember that you can use it inside ROS nodes. What we will explain here is how to link to OpenCV in your packages or nodes.
- We will also use the visualization tools already explained in *Chapter 3, Debugging and Visualization*. We will refer to the `image_view` package and the visualization nodes for monocular and stereo vision. Special attention is given to the stereo one, since it allows configuring the disparity image algorithm correctly. Then, in **rviz**, we can see the 3D point cloud obtained from the calibrated stereo pair and the fine-tuned disparity image.
- Special attention is given to the camera calibration, whose results can be integrated in a particular camera information message. Some tools help in the process of calibration, with support for different calibration patterns and the calibration optimization engine in the kernel. Furthermore, we cover stereo cameras and explain how to manage rigs of two or more cameras, with more complex setups than a binocular camera.

  Note that in *Chapter 3, Debugging and Visualization* we briefly show how to calibrate cameras, but here we go a step further explaining how everything is integrated in the system. For instance, stereo vision will let you obtain depth information from the world, up to some extent and with certain conditions. Hence, we will see how to inspect that information as point clouds and how to get the best quality and setup for our camera.

- ROS comes with an image pipeline that simplifies the process of converting RAW images acquired by the camera into monochrome (grayscale) and color images. This sometimes means to de-Bayer the RAW image, if it is codified as a Bayer pattern, which is common in high-quality FireWire cameras. If the camera has been calibrated, the calibration information is used to rectify the images. By rectifying images we mean to correct the distortion using the distortion coefficients computed during the calibration process. Finally, for stereo images, since we have the baseline between the left and right cameras, we can compute the disparity image that allows us to obtain depth information and a 3D point cloud once it has been fine-tuned. Here we will give some advice on how to tune it, which may be quite difficult for low-quality cameras and sometimes requires good calibration results beforehand.
- There are also some stacks or just packages that provide very powerful algorithms to perform interesting robotics tasks in ROS. We will enumerate some of them and will also show an example for visual odometry. In particular, this chapter will finish with a tutorial to set up and run the **viso2_ros** wrapper of the **libviso2** visual odometry library, using a stereo pair built with two cheap webcams attached to a bar. Other visual odometry libraries will be mentioned, for example, **fovis**, as well as some advice to start working with them, and how to improve the results with RGB-D sensors, such as Kinect, or even sensor fusion or additional information in the case of monocular vision.

> Note that most of these tools are quite recent and you will need ROS Groovy and the catkin-style build system, as we will be explaining in the *Performing visual odometry with viso2* section. Since the visual odometry requires good cameras, we have also included a demo using some images available online.

Connecting and running the camera

We are going to start from the very beginning. The first step we must accomplish is connecting our camera to our computer, running the driver, and checking the images it acquires in ROS. Before we go into ROS, it is always a good idea to use external tools to check that the camera is actually recognized by our system, which in our case is an **Ubuntu distro**. We will start with FireWire cameras, since they are better supported in ROS, and later we will see USB ones.

FireWire IEEE1394 cameras

Connect your camera to the computer, which should have a FireWire IEEE1394a or IEEE1394b slot (you will probably need an IEEE1394 board or a laptop such as a Mac). Then, in Ubuntu, you only need **Coriander** to check that the camera is recognized and working. If not already installed, just install Coriander. Then run it (in the older Ubuntu distros, you may have to run it as `sudo`):

```
coriander
```

It will automatically detect all your FireWire cameras, as shown in the following screenshot:

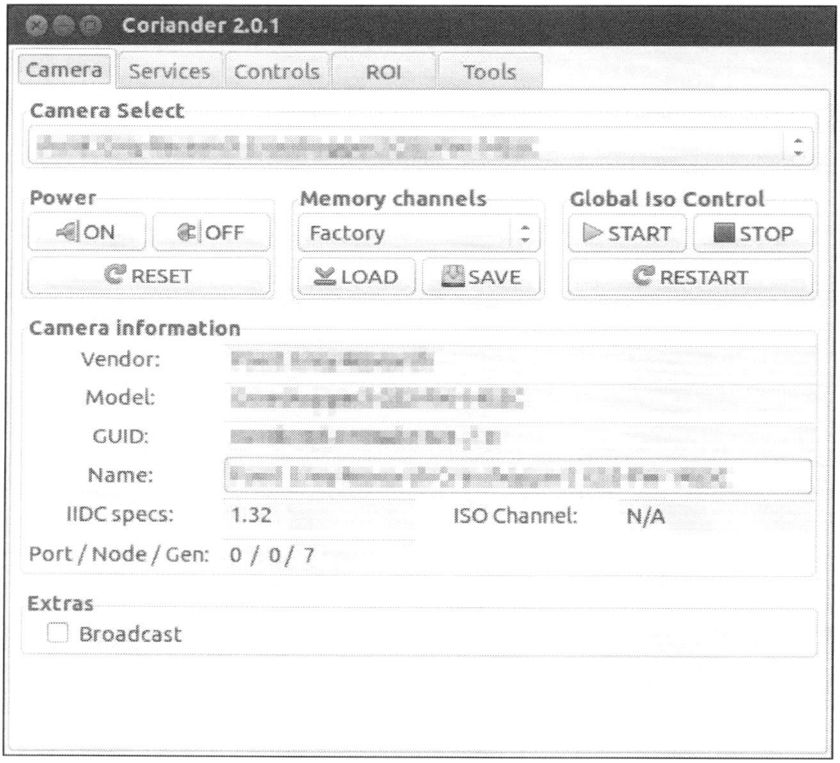

The great thing about Coriander is that it also allows us to view the image and configure the camera. Indeed, our advice is to use Coriander's camera configuration interface and then take those values into ROS, as we will see later. The advantage of this approach is that Coriander gives us the dimensional values of some parameters and in ROS there are some parameters that sometimes fail to be set, for example, `gamma`, and they must be set beforehand in Coriander as a workaround.

Now that we know the camera is working, we can close Coriander and run the ROS driver using the following command:

```
rosrun camera1394 camera1394_node
```

Just have `roscore` running and run the preceding command. It will run the first camera on the bus, but note that you can select the camera by its GUID, which you can see in Coriander's GUI. In order to select the camera by its GUID and set its working mode and all its parameters, we are going to use a file because there are so many to pass them by the command line. This is, therefore, the way to do it in a maintainable manner, which can be integrated in a `.launch` file as well.

The FireWire camera parameters supported are listed and assigned certain values in the `params/firewire_camera/format7_mode0.yaml` file, shown as follows:

```
guid: 00b09d0100ab1324       # (defaults to first camera on bus)
iso_speed: 800  # IEEE1394b
video_mode: format7_mode0  # 1384x1036 @ 30fps bayer pattern
# Note that frame_rate is overwritten by frame_rate_feature; some
useful values:
# 21fps (480)
frame_rate: 21 # max fps (Hz)
auto_frame_rate_feature: 3 # Manual (3)
frame_rate_feature: 480
format7_color_coding: raw8  # for bayer
bayer_pattern: rggb
bayer_method: HQ
auto_brightness: 3 # Manual (3)
brightness: 0
auto_exposure: 3 # Manual (3)
exposure: 350
auto_gain: 3 # Manual (3)
gain: 700
# We cannot set gamma manually in ROS, so we switch it off
auto_gamma: 0 # Off (0)
#gamma: 1024    # gamma 1
auto_saturation: 3 # Manual (3)
saturation: 1000
auto_sharpness: 3 # Manual (3)
sharpness: 1000
auto_shutter: 3 # Manual (3)
#shutter: 1000 # = 10ms
shutter: 1512 # = 20ms (1/50Hz), max. in 30fps
auto_white_balance: 3 # Manual (3)
```

```
white_balance_BU: 820
white_balance_RV: 520
frame_id: firewire_camera
camera_info_url: package://chapter6_tutorials/calibration/firewire_
camera/calibration_firewire_camera.yaml
```

The values must be obtained just by watching the images acquired, for example, in Coriander, and setting the values that give better images. The GUID parameter is used to select our camera, which is a unique value. You should usually set the shutter speed to a frequency equal to or a multiple of the electric light (for instance 50 Hz or 60 Hz, in Europe or USA, respectively) you have in the room. This can be done with classical light bulbs and fluorescents as well. If outside with sunlight, you only have to worry about setting a value that gives you a light image. You can also put a high gain, but it will introduce noise. However, in general it is better to have such a salt-and-pepper noise than a low shutter speed (to receive most light), because with a low shutter speed we will encounter motion blur and most algorithms perform badly with that. As you will see, the configuration depends on the lighting conditions in the environment and you may have to adapt it to them. This is quite easy using the ROS Fuerte `reconfigure_gui` interface:

```
rosrun dynamic_reconfigure dynamic_reconfigure_gui /camera
```

In the following screenshot we see a particular configuration in Coriander and the output image:

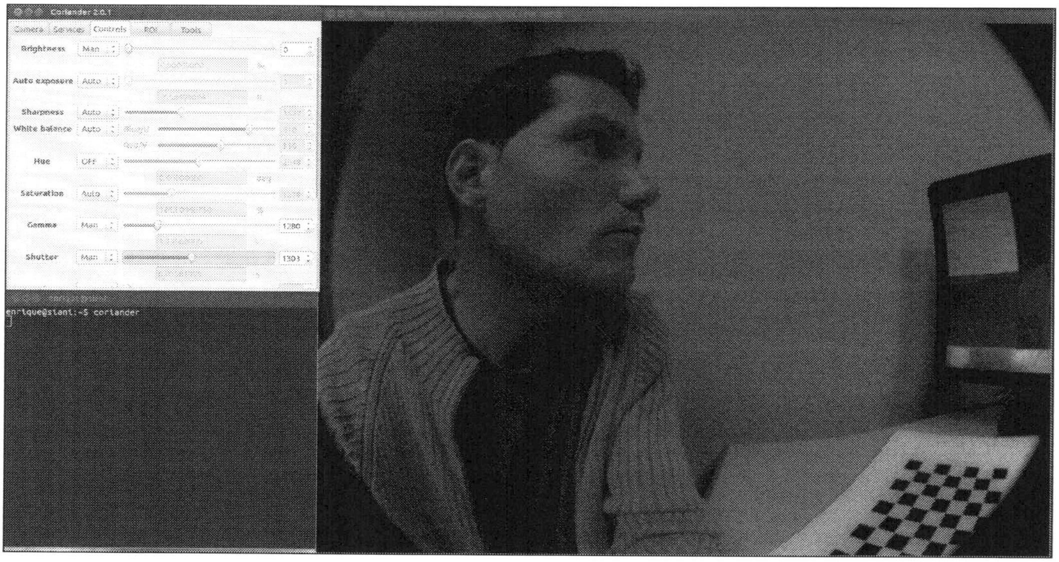

In the following screenshot we see a different configuration with a better exposure of the resulting image acquired:

We assume the camera's namespace is `/camera`. Then, we can change all the parameters, which are specified in the `camera1394` CFG file, as we have seen in *Chapter 3, Debugging and Visualization*. Here, for your convenience, you can create a `.launch` file and name it `firewire_camera.launch` (with the following code) and place it in the `launch` folder:

```
<launch>
  <!-- Arguments -->
  <!-- Show video output (both RAW and rectified) -->
  <arg name="view" default="false"/>
  <!-- Camera params (config) -->
  <arg name="params" default="$(find chapter6_tutorials)/params/
firewire_camera/format7_mode0.yaml"/>

  <!-- Camera driver -->
  <node pkg="camera1394" type="camera1394_node" name="camera1394_
node">
    <rosparam file="$(arg params)"/>
  </node>

  <!-- Camera image processing (color + rectification) -->
```

```
  <node ns="camera" pkg="image_proc" type="image_proc" name="image_
proc"/>

  <!-- Show video output -->
  <group if="$(arg view)">
    <!-- Image viewer (non-rectified image) -->
    <node pkg="image_view" type="image_view" name="non_rectified_
image">
      <remap from="image" to="camera/image_color"/>
    </node>

    <!-- Image viewer (rectified image) -->
    <node pkg="image_view" type="image_view" name="rectified_image">
      <remap from="image" to="camera/image_rect_color"/>
    </node>
  </group>
</launch>
```

It starts the `camera1394` driver with the parameters shown thus far. Then, it also runs the image pipeline we will see further in the code in order to obtain the color-rectified images using the de-Bayering algorithm and the calibration parameters (once the camera has been calibrated). Finally, we have a conditional group to visualize the color and color-rectified images using `image_view`. We use groups intensively in other `.launch` files that come with the code accompanying this book.

To sum it up, in order to run your FireWire camera in ROS, and view the images once you have set its GUID in the parameters file, just run the following command:

```
roslaunch chapter6_tutorials firewire_camera.launch view:=true
```

Then, you can also configure it dynamically with `reconfigure_gui`.

USB cameras

Now we are going to do the same with USB cameras. The only problem is that, surprisingly, they are not inherently supported by ROS. First of all, once you connect your camera to the computer, test it with any chatting or video meeting program, for example, Skype or Cheese. The camera resource should appear in `/dev/videoX`, where `X` should be a number starting from 0 (that may be your internal webcam if you are using a laptop).

There are two main options that deserve to be mentioned as possible USB camera drivers for ROS. First, we have `usb_cam`. To install it, run the following command:

```
svn co http://svn.code.sf.net/p/bosch-ros-pkg/code/trunk/stacks/bosch_
drivers/usb_cam
rosstack profile && rospack profile
roscd usb_cam
rosmake
```

After that is done, you can run the following command:

```
roslaunch chapter6_tutorials usb_cam.launch view:=true
```

It just runs `rosrun usb_cam usb_cam_node` and also shows the camera images with `image_view`, so you should see something like the following screenshot. It has the RAW image of the USB camera, which is already in color:

Similarly, another good option is `gscam`, which is installed as follows:

```
$ svn co http://brown-ros-pkg.googlecode.com/svn/trunk/experimental/gscam
$ echo 'include $(shell rospack find mk)/cmake.mk' > Makefile
rosmake
```

Note that we had to change the `Makefile` file because it was wrong (this is the content of all `Makefile` files inside ROS packages, but it seems that this `Makefile` file was ignored by an ignoring rule in the repository of the software). Then, just run this command:

```
roslaunch chapter6_tuturials gscam.launch view:=true
```

As for `usb_cam`, this `.launch` file runs `rosrun gscam gscam`, and also sets some camera parameters. It also visualizes the camera images with `image_view`, as shown in the following screenshot:

The parameters required by `gscam` are (see `params/gscam/logitech.yaml`) as follows:

```
gscam_config: v4l2src device=/dev/video0 ! video/x-raw-rgb,framerate=30/1
! ffmpegcolorspace

frame_id: gscam

camera_info_url: package://chapter6_tutorials/calibration/gscam/
calibration_gscam.yaml
```

The `gscam_config` command invokes the `v4l2src` command with the appropriate arguments to run the camera. The rest of the parameters will be useful once the camera is calibrated and used in an ROS image pipeline.

Making your own USB camera driver with OpenCV

Although we have the two options previously mentioned, this book comes with its own USB camera driver, implemented on top of OpenCV, using the `cv::VideoCapture` class. It runs the camera and also allows changing some of its parameters, as long as they are supported. This is very simple and it allows setting the calibration information in the same way as with FireWire cameras. Indeed, with `usb_cam` this is not possible because the `CameraInfo` message is not available. Hence, `params/usb_cam/logitech.yaml the camera_info_url` is not set in the parameters file. With regard to `gscam`, we will have more control.

We can change the camera configuration and also see how to publish the camera images and information in ROS from scratch. In order to implement a camera driver using OpenCV, we have two options regarding the way in which to read images from the camera. First, we can do a polling according to a given number of frames per second (FPS). Second, we can set a timer for the period of such FPS and during the callback of the timer we perform the actual reading. Depending on the FPS, our solution may be better than the others in terms of CPU consumption. Anyway, note that the polling is not active, since the OpenCV reading function is blocked, and other processes can take the CPU until an image is ready. In general, for fast FPS, it is better to use polling, so we do not incur a time penalty for using the timer and its callback. For a low FPS, the timer should be similar to the polling, and the code is way cleaner. We invite the reader to compare both implementations in the `src/camera_polling.cpp` and `src/camera_timer.cpp` files. For the sake of space, here we show the timer-based approach. Indeed, the final driver, in `src/camera.cpp`, uses a timer. Note that the final driver also includes the camera information management, which we will see in the following section.

Creating the USB camera driver package

In the `manifest.xml` file, we must set the dependency with OpenCV, the ROS `Image` message libraries, and the related packages. They are as follows:

```
<depend package="sensor_msgs"/>
<depend package="opencv2"/>
<depend package="cv_bridge"/>
<depend package="image_transport"/>
```

Consequently, in `src/camera_timer.cpp` we have the following headers:

```
#include <ros/ros.h>
#include <image_transport/image_transport.h>
#include <cv_bridge/cv_bridge.h>
#include <sensor_msgs/image_encodings.h>
#include <opencv2/highgui/highgui.hpp>
```

The `image_transport` API allows publishing the images using several transport formats, which can be compressed images and different codecs being done seamlessly for the user based on plugins installed in the ROS system, for example, compressed Theora. The `cv_bridge` element is used to convert from OpenCV Image to the ROS `Image` message, for which we may need the image encodings of `sensor_msgs`, in case of grayscale/color conversion. Finally, we need the `highgui` API of OpenCV (`opencv2`), in order to use `cv::VideoCapture`.

Here we will explain the main parts of the code in `src/camera_timer.cpp`, which have a class that implements the camera driver. Its attributes are as follows:

```
ros::NodeHandle nh;
image_transport::ImageTransport it;
image_transport::Publisher pub_image_raw;

cv::VideoCapture camera;
cv::Mat image;
cv_bridge::CvImagePtr frame;

ros::Timer timer;

int camera_index;
int fps;
```

Using the ImageTransport API to publish the camera frames

As usual, we need the node handle. Then, we need an `ImageTransport` object, which is used to send the images in all available formats in a seamless way. In the code we only need to use the publisher, but note that it must be the implementation of the `image_transport` library, and not the common `ros::Publisher` for an Image message.

Then we have the OpenCV stuff to capture images/frames. In the case of the frame, we directly use the `cv_brigde` frame, which is a `CvImagePtr`, since we can access the image field it has.

Finally, we have the timer and the basic camera parameters for a driver to work. This is the most basic driver possible. These parameters are the camera index, that is, the number for the `/dev/videoX` device, for example, `0` for `/dev/video0`. The camera index is passed to `cv::VideoCapture`. Finally, the FPS is used to set the camera (if possible, since some cameras do not support this) and the timer. Here we use an `int` value, but it will be a `double` value in the final version of `src/camera.cpp`.

The driver uses the class constructor for the setup or initialization of the node, the camera, and the timer:

```
nh.param<int>( "camera_index", camera_index, DEFAULT_CAMERA_INDEX );

if ( not camera.isOpened() )
```

```
{
    ROS_ERROR_STREAM( "Failed to open camera device!" );
    ros::shutdown();
}

nh.param<int>( "fps", fps, DEFAULT_FPS );
ros::Duration period = ros::Duration( 1. / fps );

pub_image_raw = it.advertise( "image_raw", 1 );

frame = boost::make_shared< cv_bridge::CvImage >();
frame->encoding = sensor_msgs::image_encodings::BGR8;

timer = nh.createTimer( period, &CameraDriver::capture, this );
```

First, we open the camera, or abort if it is not open. Note that we must do this in the attribute constructor:

```
camera( camera_index )
```

Here `camera_index` is passed as parameter.

Then, we read the `fps` parameter and compute the timer period, which is used to create the timer and set the capture callback at the end. We advertise the image publisher using the ImageTransport API, for `image_raw` (RAW images), and initialize the `frame` variable.

The capture callback reads and publishes the images as follows:

```
camera >> frame->image;
if( not frame->image.empty() )
{
    frame->header.stamp = ros::Time::now();
    pub_image_raw.publish( frame->toImageMsg() );
}
```

It captures the images, check whether a frame was actually captured, and in that case, sets the timestamp and publishes the image, which is then converted to an ROS Image.

You can run this node with:

```
rosrun chapter6_tutorials camera_timer camera_index:=0 fps:=15
```

This will open the `/dev/video0` camera at 15 FPS.

Then you can use `image_view` to see the images. Similarly, for the polling implementation, you have a `.launch` file, so you can run the following command:

```
roslaunch chapter6_tutorials camera_polling.launch camera_index:=0
fps:=15 view:=true
```

Now, you will see the `/camera/image_raw` topic images.

For the timer implementation, we have the `camera.launch` file that runs the final version and provides more options, which we will see throughout this chapter. The main contributions of the final version are the support for dynamic reconfiguration parameters, that is, it provides the camera information, which includes the camera calibration. We are going to show how to do this in brief and we advise the reader to check the source code for a more detailed view as well as the entire coding.

Similar to FireWire cameras, we can provide support for dynamic reconfiguration of the camera parameters. However, most USB cameras do not support the changing of some parameters. What we do is expose all OpenCV supported parameters and warn the user in case of an error (or disable some of them). The configuration file is in `cfg/Camera.cfg`. Check it for details. It supports these parameters:

- `camera_index`: This is used to select the `/dev/videoX` device.
- `frame_width` and `frame_height`: These provide the image resolution.
- `fps`: This provides the camera's FPS value.
- `fourcc`: This specifies the camera pixels in the FOURCC format (see `http://www.fourcc.org`, although they are typically YUYV or MJPG, but they fail to change in most USB cameras with OpenCV).
- `brightness`, `contrast`, `saturation`, and `hue`: These values set the camera's properties. In digital cameras, this is done using software to the acquisition process in the sensor or just to the resulting image.
- `gain`: This parameter sets the gain of the **Analog-Digital converter** (**ADC**) of the sensor. It introduces the salt-and-pepper noise into the image, but increases the lightness in dark environments.
- `exposure`: This determines the exposure of the images, which sets the lightness of the images, usually by configuring the gain and shutter speed (in low-cost cameras this is just the integration time of the light that enters the sensor).
- `frame_id`: This specifies the camera frame, and is useful if we use it for navigation, as we will see in the visual odometry section.
- `camera_info_url`: This is the path to the camera information, which is basically its calibration.

Then, in the driver, we use a dynamic reconfigure server using:

```
#include <dynamic_reconfigure/server.h>
```

We set a callback in the constructor:

```
server.setCallback( boost::bind( &CameraDriver::reconfig, this, _1, _2
) );
```

And the callback reconfigures the camera. We even allow changing the camera, and stopping the current one. Then we use OpenCV's `cv::VideoCapture` class to set the camera properties, which are part of the previously mentioned parameters. We will see this in the case of the `frame_width` parameter as an example:

```
newconfig.frame_width = setProperty( camera, CV_CAP_PROP_FRAME_WIDTH ,
newconfig.frame_width );
```

It relies on a private method named `setProperty` that calls the `set` method of `cv::VideoCapture` and controls the cases where it fails to send an ROS warning message.

The FPS is changed in the timer itself and usually cannot be modified in the camera as the other properties.

Finally, it is important to note that all this reconfiguration is done within a locked mutex to avoid acquiring any images while reconfiguring the driver.

In order to set the camera information, ROS has a `camera_info_manager` library, which helps us to do so. In short, we use:

```
#include <camera_info_manager/camera_info_manager.h>
```

And we use it to obtain the `CameraInfo` message. Now, in the capture callback of the timer, we can use `image_transport::CameraPublisher` (and not just for the images). The code is as follows:

```
camera >> frame->image;
if( not frame->image.empty() )
{
  frame->header.stamp = ros::Time::now();

  *camera_info = camera_info_manager.getCameraInfo();
  camera_info->header = frame->header;

  camera_pub.publish( frame->toImageMsg(), camera_info );
}
```

This is run within the preceding mutex for the reconfiguration method. Now, we do this for the first version of the driver but also retrieve the camera information from the manager, which is set to the node handler, the camera name, and the `camera_info_url` parameter, in the reconfiguration method (which is always called once on loading). Then, we publish both the image/frame (an ROS Image) and the `CameraImage` messages.

In order to use this driver, just run the following command:

```
roslaunch chapter6_tutorials camera.launch view:=true
```

It will use the `params/camera/webcam.yaml` parameters as default, which sets all the dynamic reconfiguration parameters seen thus far.

You can check whether the camera is working with `rostopic list`, `rostopic hz /camera/image_raw`, and also with `image_view`.

With the implementation of this driver we have used all the resources available in ROS to work with cameras, images, and computer vision. In the following sections we will separately explain each of them, for the sake of clarity.

Dealing with OpenCV and ROS images using cv_bridge

Assume we have an OpenCV image, that is, a `cv::Mat image`. We need the `cv_bridge` library to convert it into a ROS `Image` message and publish it. We have the option to share or copy the image, with `CvShare` or `CvCopy`, respectively. However, if possible, it is easier to use the OpenCV image field inside the `CvImage` class provided by `cv_bridge`. That is exactly what we do in the camera driver, as a pointer:

```
cv_bridge::CvImagePtr frame;
```

Being a pointer, we initialize it this way:

```
frame = boost::make_shared< cv_bridge::CvImage >();
```

And if we know the image encoding beforehand:

```
frame->encoding = sensor_msgs::image_encodings::BGR8;
```

Later, we set the OpenCV image at some point, for example, capturing it from a camera:

```
camera >> frame->image;
```

It is also common to set the timestamp of the message at this point:

```
frame->header.stamp = ros::Time::now();
```

Now we only have to publish it. To do so, we need a publisher and it must use the image transport API of ROS. This is shown in the following section.

Publishing images with ImageTransport

We can just publish single images with `ros::Publisher`, but it is better to use the `image_transport` publishers. It can publish both simple images or images with their corresponding camera information. That is exactly what we do for the previously mentioned camera driver. The ImageTransport API is useful to provide different transport formats in a seamless way. The images you publish actually appear in several topics. Apart from the basic, uncompressed one, you will see a compressed one or even more. The number of supported transports depends on the plugins installed in your ROS environment; you will usually have the `compressed` and `theora` transports. You can see this with a simple `rostopic call` command.

In your code you need the node handle to create the image transport and then the publisher. In this example we will use a simple image publisher. Please check the final USB camera driver for the `CameraPublisher` usage:

```
ros::NodeHandle nh;
image_transport::ImageTransport it;
image_transport::Publisher pub_image_raw;
```

The node handle and the image transport are constructed with the following code (in the attribute constructors of a class):

```
nh( "~" ),
it( nh )
```

Then, the publisher is created this way, for an `image_raw` topic, within the `node` namespace:

```
pub_image_raw = it.advertise( "image_raw", 1 );
```

Hence, now the `frame` attribute shown in the previous section can be published using the following code statement:

```
pub_image_raw.publish( frame->toImageMsg() );
```

Using OpenCV in ROS

ROS Fuerte comes with one of the latest stable versions of OpenCV; in newer ROS distributions it simply depends on the Debian package of the OpenCV library installed in your system. In order to use it in our nodes, include the following code snippet in your `manifest.xml` file:

```
<depend package="opencv2"/>
```

In the `CMakeLists.xml` file we only have to put a line to build our node, that is, nothing must be done regarding the OpenCV library.

In our node's `.cpp` file we include any of the OpenCV libraries we need. For example, for the `highgui.hpp` file we use the following statement:

```
#include <opencv2/highgui/highgui.hpp>
```

Now, you can use any of the OpenCV API classes, functions, and many more code elements, in your code, as a regular. Just use its namespace, `cv`, and follow any OpenCV tutorial if you are new to OpenCV. Note that this book is not about OpenCV, just how to do computer vision inside ROS. Then, compile everything with `rosmake` as usual for an ROS package.

Visualizing the camera input images

In *Chapter 3, Debugging and Visualization*, we explained how to visualize any image published in the ROS framework, using the `image_view` topic of the `image_view` package:

```
rosrun image_view image_view image:=/camera/image_raw
```

What is important here is the fact that using the image transport we can select different topics to see the images, using compressed formats if required. Also, in the case of stereo vision, as we will see later, we can use `rviz` to see the point cloud obtained with the disparity image.

How to calibrate the camera

Most cameras, especially wide angular ones, exhibit a large distortion. We can model such distortion as radial or tangential and compute the coefficients of that model using calibration algorithms. The camera calibration algorithms also allow us to obtain a calibration matrix that contains the focal distance and principal point of the lens, and hence provides a way to measure distances in meters in the world using the images acquired. In the case of stereo vision, it is also possible to retrieve depth information, that is, the distance of the pixels to the camera, as we will see later. Consequently, we will have 3D information of the world.

The calibration is done by showing several views of a known image named **calibration pattern**, which is typically a chessboard. It can also be an array of circles, or an asymmetric pattern of circles. Note that circles are seen as ellipses by the cameras for skew views. A detection algorithm obtains the inner corner point of the cells on the chessboard and uses them to estimate the camera's intrinsic and extrinsic parameters. In short, the extrinsic parameters are the poses of the camera, or in other words, the poses of the pattern with regard to the camera, if we left the camera in a fixed position. What we want are the intrinsic parameters, because they do not change and can be used later for the camera at any pose, and allow measuring distances in the images and correcting the image distortion, that is, rectifying the image.

With our camera driver running, we can use the calibration tool of ROS to calibrate it. It is important that the camera driver provides `CameraInfo` messages and has the `camera_info_set` service, which allows setting the path to the calibration results file. Later, this calibration information is always loaded by the image pipeline when using the camera. One camera driver that satisfies these prerequisites is the `camera1394` driver for FireWire cameras. In order to calibrate your FireWire camera, just run the following command:

```
roslaunch chapter6_tutorials calibration_firewire_camera_chessboard.launch
```

This will open a GUI that automatically selects the views of our calibration pattern, and it provides some bars to inform how each "axis" is covered by the views retrieved. It comprises of the *x* and *y* axes, meaning it shows the patterns close to each extreme of these axes in the image plane, that is, the horizontal and vertical axis, respectively. Then, the scale goes from close to far (up to that distance at which the detection works). Finally, skew requires views of the pattern tilt in both the *x* and *y* axes. The three buttons below these bars are disabled by default, as shown in the following screenshot:

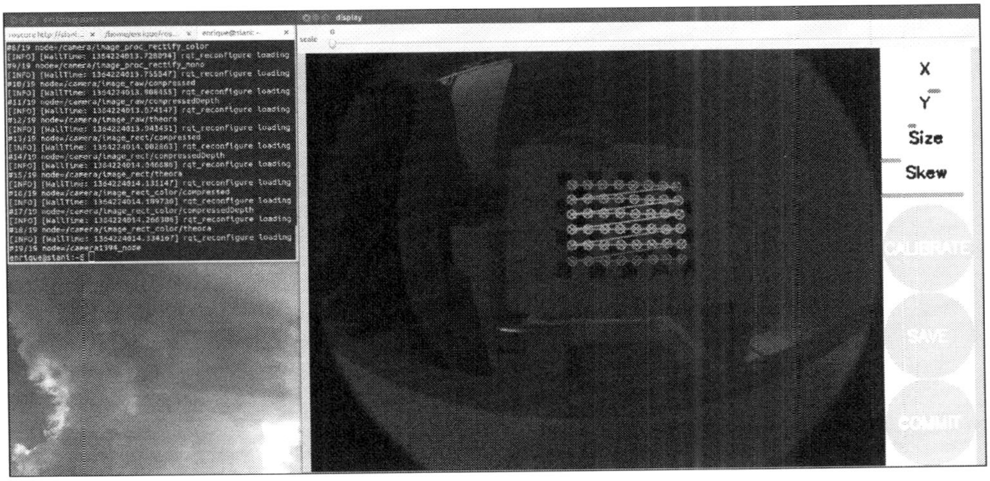

You will see the points detected in the pattern overlay every time the detector can do it. The views are automatically selected to cover a representative number of different views, so you can make the bars go green from one side to the other, following the instructions given in a moment. In theory, two views are enough, but in practice around 10 are usually needed. In fact, this interface captures even more (30 to 40). You should avoid fast movements because blurry images are bad for detection. Once the tool has enough views, it will allow you to calibrate, that is, to start the optimizer that solves the system of the pinhole camera model given the points detected in the calibration pattern views.

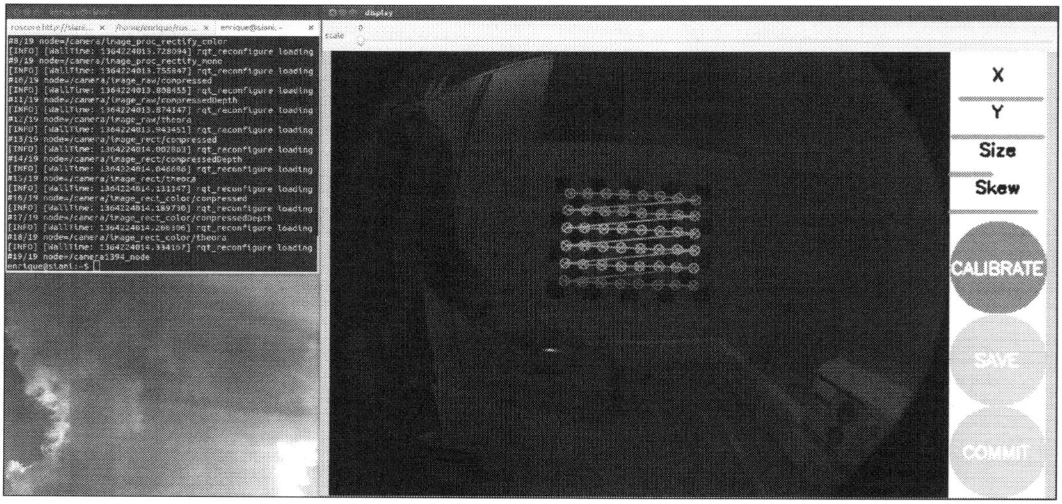

Then, you can save the calibration data and commit the calibration results to the camera. For this, it uses the `camera_info_set` service to commit the calibration to the camera, so later it is detected automatically by the ROS image pipeline.

For the `.launch` file provided for the calibration, simply use the `cameracalibrator.py` file of the ROS package `camera_calibration`, using the following node:

```
<node pkg="camera_calibration" type="cameracalibrator.py"
      name="cameracalibrator" args="--size 8x6 --square 0.030"
      output="screen">
  <remap from="image" to="camera/image_color" />
  <remap from="camera" to="camera" />
</node>
```

It also uses the image pipeline but it is not required. In fact, instead of the `image_color` topic, we could use the `image_raw` one.

Once you have saved the calibration (using the **Save** button), a file is created in your `temp` directory. It contains the calibration pattern views used for the calibration. You can find it at `/tmp/calibrationdata.tar.gz`. The ones used for the calibration in the book, can be found in the `calibration` directory, with `firewire_camera` being the first case for the FireWire camera. Similarly, on the terminal (the `stdout` output), you will see information regarding the views taken and the calibration results. The ones obtained for the book are in the same folder as the calibration data. The calibration results can also be consulted in the `ost.txt` file inside the `calibrationdata.tar.gz` folder. Anyway, remember that after the commit, the calibration file is updated with the calibration matrix and the coefficients of the distortion model. A good way of doing this consists of creating a dummy calibration file before the calibration. In our package, that file is at `calibration/firewire_camera/calibration_firewire_camera.yaml`, which is accessed by the parameters file:

```
camera_info_url: package://chapter6_tutorials/calibration/firewire_
camera/calibration_firewire_camera.yaml
```

Now, we can run our camera again with the image pipeline, and the rectified images will have the distortion corrected as a clear sign that the camera is calibrated correctly. We will see this later for the image pipeline.

For more details on the calibration formulas, since ROS uses the Zhang calibration method implemented in OpenCV, our advice is to consult its documentation. However, we think it is enough to have the user knowledge provided here.

Finally, you can also play with different calibration patterns using the following `.launch` files for circles and asymmetric circles (see `http://docs.opencv.org/trunk/_downloads/acircles_pattern.png`), meant for FireWire cameras:

```
roslaunch chapter6_tutorials calibration_firewire_camera_circles.launch
roslaunch chapter6_tutorials calibration_firewire_camera_acircles.launch
```

You can also use the multiple chessboard patterns for a single calibration, using patterns of different size. However, we think it is enough and simple to use a single chessboard pattern printed with good quality. Indeed, for the USB camera driver we only use that.

In the case of a USB camera driver, we have a more powerful `.launch` file that integrates the camera calibration node. There is also a standalone one like the one for FireWire cameras, though. Hence, in order to calibrate your camera, just run the following command:

```
roslaunch chapter6_tutorials camera.launch calibrate:=true
```

In the following screenshot you will see the steps of the calibration process in the GUI, identical to the one with FireWire cameras. That means we have an operating `camera_info_set` service:

The preceding screenshot shows the instant when enough views of the pattern have been acquired, so the calibrate button is enabled because now it is possible to solve the calibration problem. The following screenshot shows the end of the calibration process and thus allows saving the calibration data as well as committing it to the camera configuration files so that later we do not have to set anything up.

> **Downloading the example code**
>
> You can download the example code files for all Packt books you have purchased from your account at `http://www.packtpub.com`. If you purchased this book elsewhere, you can visit `http://www.packtpub.com/support` and register to have the files e-mailed directly to you. You can also download these code files from `https://github.com/AaronMR/Learning_ROS_for_Robotics_Programming`.

Stereo calibration

The following section will speak of working with stereo cameras. One option is to run two monocular camera nodes, but in general it is better to consider the whole stereo pair as a single sensor, because the images must be synchronized. In ROS there is no driver for FireWire stereo cameras, but we can use an extension to stereo provided here:

```
git clone git://github.com/srv/camera1394stereo.git
```

However, FireWire stereo pairs are quite expensive. For this reason, we provide a stereo camera driver for USB cameras. We use the Logitech C120 USB webcam, which is very cheap. It is also noisy, but we will see that we can do great things with it after we calibrate it. It is important that in the stereo pair the cameras are similar, but you can try different cameras as well. Our setup for the cameras is shown in the following images. You only need two cameras pointing in the same plane.

We have a baseline of approximately 12 cm, which will also be computed in the stereo calibration process. As you can see, you only need a bar to attach the cameras to, with zip ties, for example:

A closer and frontal view shows the camera lenses, which can be regulated manually in focus. Note that you should set it to some appropriate distance (1 to 2 meters used for the baseline should be good in the most general case):

Now, connect the cameras to your USB slots. It is good practice to connect the left camera first and then the right one. This way, they are assigned the `/dev/video0` and `/dev/video1` devices, or the `1` and `2` if `0` is already taken.

Then, you can test each camera individually, as we would do for a single camera. Some tools you will find useful are `v4l-utils` and `qv4l2` used in the following manner:

```
sudo apt-get install v4l-utils qv4l2
```

You may experience this problem:

```
libv4l2: error turning on stream: No space left on device
```

This happens because you must connect each camera to a different USB controller. Note that some USB slots are managed by the same controller, and hence it cannot deal with the bandwidth of more than a single camera. If you only have a USB controller, there are other options you can try. First, try to use a compressed pixel format, such as MJPG, in both cameras. You can check whether it is supported by your camera using:

```
v4l2-ctl -d /dev/video2 --list-formats
ioctl: VIDIOC_ENUM_FMT
Index       : 0
Type        : Video Capture
Pixel Format: 'YUYV'
Name        : YUV 4:2:2 (YUYV)

Index       : 1
Type        : Video Capture
Pixel Format: 'MJPG' (compressed)
Name        : MJPEG
```

If MJPG is supported, we can use more than one camera in the same USB controller. Otherwise, with uncompressed pixel formats, we must use different USB controllers, or reduce the resolution to 320 x 240 or below. Similarly, with the GUI of `qv4l2` you can check this and test your camera. You can also check whether it is possible to set the desired pixel format. Sometimes this is not possible. Indeed, it did not work for our USB cameras using the OpenCV `set` method, so we use a USB slot managed by a different USB controller.

The USB stereo camera driver that comes with this book is based on the USB camera driver discussed thus far (in the initial sections). Basically, it extends its support to camera publishers, which sends the left and right image and the camera info as well. You can run it and view the images with this command:

```
roslaunch chapter6_tutorials camera_stereo.launch view:=true
```

This command should work correctly if you only connect two cameras to your computer (and it already has a camera integrated), so the cameras' IDs are `1` and `2`. If you need to change them, you must edit the `params/camera_stereo/logitech_c120.yaml` file (which is used by default in the previous `.launch` file) and set the appropriate IDs in these fields:

```
camera_index_left: 1
camera_index_right: 2
```

This also shows the disparity image of left and right cameras, which will be useful and discussed later, once the cameras are calibrated and used within the ROS image pipeline. In order to calibrate, just run the following command:

```
roslaunch chapter6_tutorials camera_stereo.launch calibrate:=true
```

You will see this GUI, similar to the one for monocular cameras:

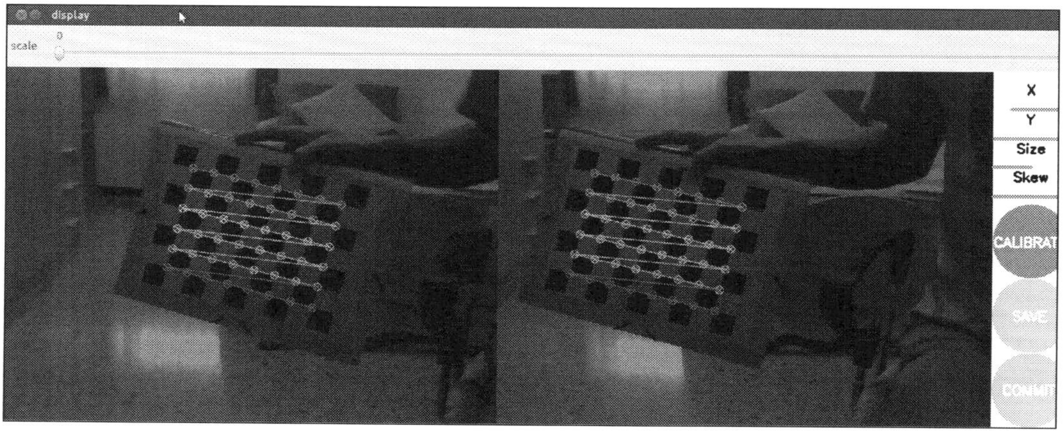

At the time of the preceding screenshot, we have shown enough views to start the calibration. Note that the calibration pattern must be detected by both cameras simultaneously to be included for the calibration optimization step. Depending on the setup this may be quite tricky, so you should put the pattern at the appropriate distance from the camera. You can see the setup used for the calibration of this book in the following image:

The calibration is done by the same `cameracalibrator.py` node as for monocular cameras. We simply pass the left and right cameras and images, so that the tool knows we are going to perform a stereo calibration. The following is the `node` element in the `.launch` file:

```
<node ns="$(arg camera)" name="cameracalibrator"
      pkg="camera_calibration" type="cameracalibrator.py"
      args="--size 8x6 --square 0.030" output="screen">
  <remap from="left" to="left/image_raw"/>
  <remap from="right" to="right/image_raw"/>
  <remap from="left_camera" to="left"/>
  <remap from="right_camera" to="right"/>
</node>
```

The result of the calibration is the same as that for monocular cameras, but in this case we have the calibration files for each camera. According to the parameters file in `params/camera_stereo/logitech_c120.yaml`, we have:

```
camera_info_url_left: package://chapter6_tutorials/calibration/camera_
stereo/${NAME}.yaml
camera_info_url_right: package://chapter6_tutorials/calibration/
camera_stereo/${NAME}.yaml
```

Here, `${NAME}` is the name of the camera, which is resolved to `logitech_c120_left` and `logitech_c120_right` for the left and right camera, respectively. After the commit of the calibration, these files are updated with the calibration of each camera. This contains the calibration matrix, the distortion model coefficients, and the rectification and projection matrix, which include the baseline, that is, the separation between each camera in the *x* axis of the image plane. In the parameters file, you can also see some values for the properties of the cameras that have been set for indoor environments with artificial light. This camera has autocorrection, so sometimes the images may be quite bad, but these values seem to work well in most cases.

The ROS image pipeline

The ROS image pipeline is run with the `image_proc` package. It provides all the conversion utilities to obtain monochrome and color images from the RAW images acquired from the camera. In the case of FireWire cameras, which may use a Bayer pattern to code the images (actually in the sensor itself), it de-Bayers them to obtain the color images. Once you have calibrated your camera, the image pipeline takes the `CameraInfo` messages, which contain the de-Bayered pattern information, and rectifies your images. Here, rectifies means to undistort the images, so it takes the coefficients of the distortion model to correct the radial and tangential distortion.

As a result, you will see more topics for your camera in its namespace. In the following screenshots, you will see the `image_raw`, `image_mono`, and `image_color` topics that show the RAW, monochrome, and color images, respectively:

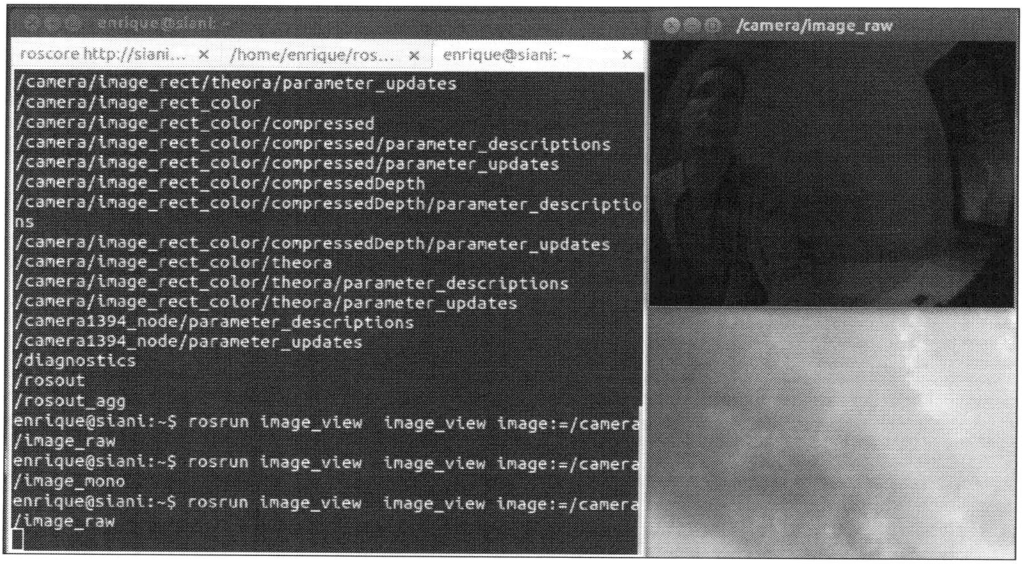

The monochrome image in the following screenshot is, in this case, the same as the RAW one, but the sensors with a Bayer pattern will be seen in the RAW image:

Finally, in the `/image_color` topic we have the image in color. Note that the process of conversion does not have to be RAW, monochrome, and color, since indeed many cameras output the color images directly. Therefore, the RAW and color images are the same, and the monochrome one is obtained by desaturating the colored one:

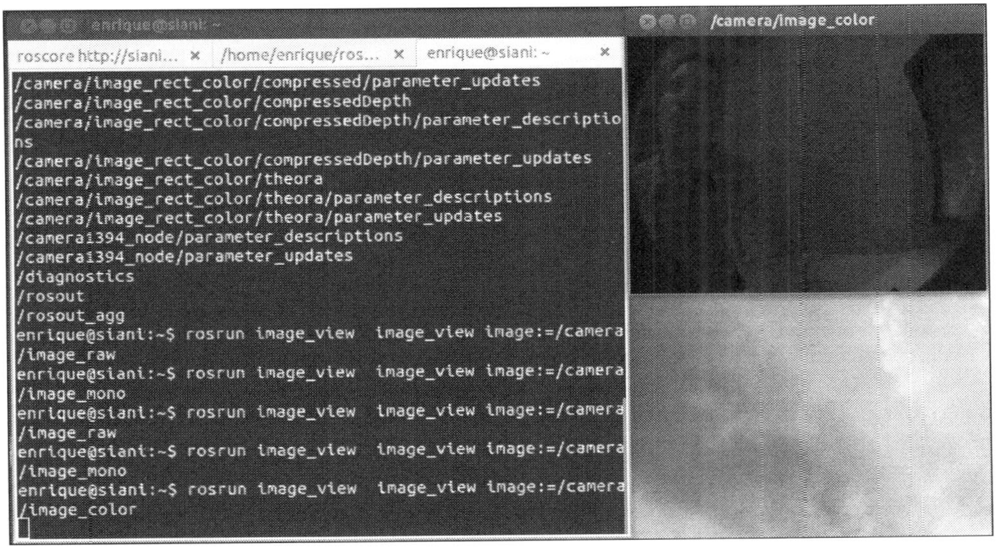

The rectified images are provided in monochrome and color, in the topics `image_rect` and `image_rect_color`. In the following screenshot we compared the uncalibrated distorted RAW images with the rectified ones. You can see the correction because the patterns shown in the images from the screenshot have straight lines only in the rectified images, particularly in the areas far from the center (principal point) of the image (sensor):

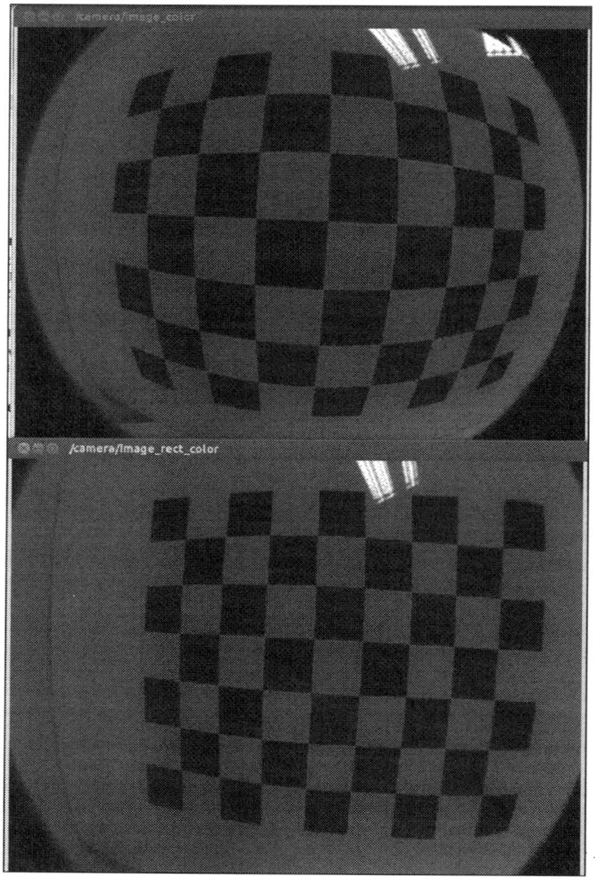

You can see all the topics available with `rostopic list` or `rxgraph`, which include the image transports as well.

You can view the `image_raw` topic of a monocular camera directly using the following command:

```
roslaunch chapter6_tutorials camera.launch view:=true
```

It can be changed to see other topics, but for these cameras, the RAW images are already in color. However, to see the rectified ones, use the `image_rect_color` topic, with `image_view`, or change the `.launch` file. The `image_proc` node is used to make all these topics available, just with the following code in the `.launch` file:

```
<node ns="$(arg camera)" pkg="image_proc" type="image_proc"
name="image_proc"/>
```

Image pipeline for stereo cameras

In the case of stereo cameras, we have the same for left and right cameras. However, there are visualization tools specific for them, because we can use the left and right images to compute and see the disparity image. An algorithm uses the stereo calibration and the texture of both images to estimate the depth of each pixel, which is the disparity image. To obtain good results, we must tune the algorithm that computes such an image. In the following screenshot we see the left, right, and disparity images, as well as `reconfiguire_gui` for `stereo_image_proc`, which is the node that builds the image pipeline for stereo images. In the `.launch` file we only need:

```
<node ns="$(arg camera)" pkg="stereo_image_proc"
    type="stereo_image_proc"
    name="stereo_image_proc" output="screen">
  <rosparam file="$(arg params_disparity)"/>
</node>
```

It requires the disparity parameters, which can be set with `reconfigure_gui` as in the following screenshot, and saved with `rosparam dump /stereo/stereo_image_proc`:

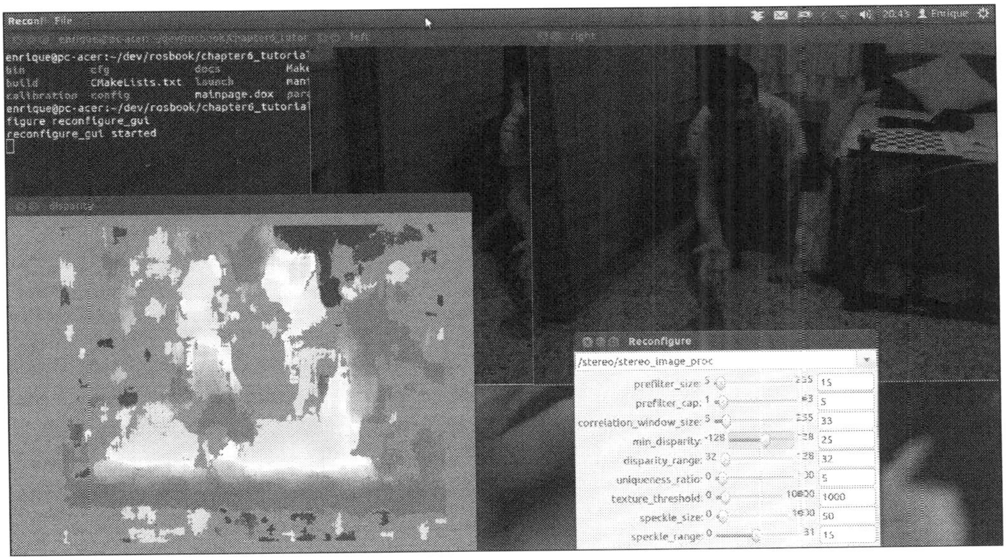

We have good values for the environment used in this book demo in the parameters file `params/camera_stereo/disparity.yaml`:

```
{correlation_window_size: 33, disparity_range: 32, min_disparity: 25, prefilter_cap: 5,
  prefilter_size: 15, speckle_range: 15, speckle_size: 50, texture_threshold: 1000,
  uniqueness_ratio: 5.0}
```

However, these parameters depend a lot on the calibration quality and the environment. You should adjust it to your experiments. It takes time and it is quite tricky, but you can follow the guidelines given on the ROS page at `http://www.ros.org/wiki/stereo_image_proc/Tutorials/ChoosingGoodStereoParameters`.

Basically, you start by setting a `disparity_range` value that allows enough blobs. You also have to set `min_disparity`, so you see areas covering the whole range of depths (from red to blue/purple). Then, you can fine-tune the result setting `speckle_size` to remove small noisy blobs. Also, modify `uniqueness_ratio` and `texture_threshold` to have larger blobs. The `correlation_window_size` parameter is also important since it affects the detection of initial blobs.

If it becomes very difficult to obtain good results, you may have to recalibrate or use better cameras for your environment and lighting conditions. You can also try in another environment or with more light. It is important that you have texture in the environment, for example, from a flat white wall you cannot find any disparity. Also, depending on the baseline, you cannot retrieve depth information very close to the camera. Similarly, for long distances from the camera, the depth estimation is less accurate. In conclusion, the distance of the baseline depends on the target application, and it is proportional to the depth that we want to measure. We use a value of 12 cm that is good for 1 to 2 meters, because later we will try visual odometry, which is usually performed at relatively large distances (> 1 meter). However, with this setup, we only have depth information one meter apart from the cameras. However, with a smaller baseline, we can obtain depth information from closer objects. This is bad for navigation because we lose far away resolution, but it is good for perception and grasping.

Regarding calibration problems, you can check your calibration results with the `cameracheck.py` node, which is integrated in both the monocular and stereo camera `.launch` files:

```
roslaunch chapter6_tutorials camera.launch view:=true check:=true
roslaunch chapter6_tutorials camera_stereo.launch view:=true check:=true
```

For the monocular camera, our calibration yields this RMS error (see more in `calibration/camera/cameracheck-stdout.log`):

```
Linearity RMS Error: 1.319 Pixels      Reprojection RMS Error: 1.278
Pixels
Linearity RMS Error: 1.542 Pixels      Reprojection RMS Error: 1.368
Pixels
Linearity RMS Error: 1.437 Pixels      Reprojection RMS Error: 1.112
Pixels
Linearity RMS Error: 1.455 Pixels      Reprojection RMS Error: 1.035
Pixels
Linearity RMS Error: 2.210 Pixels      Reprojection RMS Error: 1.584
Pixels
Linearity RMS Error: 2.604 Pixels      Reprojection RMS Error: 2.286
Pixels
Linearity RMS Error: 0.611 Pixels      Reprojection RMS Error: 0.349
Pixels
```

For the stereo camera, we have the epipolar error and the estimation of the cell size of the calibration pattern (see more in `calibration/camera_stereo/cameracheck-stdout.log`):

```
epipolar error: 0.738753 pixels   dimension: 0.033301 m
epipolar error: 1.145886 pixels   dimension: 0.033356 m
epipolar error: 1.810118 pixels   dimension: 0.033636 m
epipolar error: 2.071419 pixels   dimension: 0.033772 m
epipolar error: 2.193602 pixels   dimension: 0.033635 m
epipolar error: 2.822543 pixels   dimension: 0.033535 m
```

To obtain this result, you only have to show the calibration pattern to the camera(s). This is the reason we also pass `view:=true` to the `.launch` files. An RMS error greater than two pixels is quite large; we have something around it, but you will recall that these are very low-cost cameras. Something below one pixel error is desirable. For the stereo pair, the epipolar error should also be lower than a pixel; in our case, it is still quite large (usually greater than three pixels), but we can still do many things. Indeed, the disparity image is just a representation of the depth of each pixel, shown with the `stereo_view` node. We also have a 3D point cloud that we can see texturized in `rviz`. We will see this for the following demos, doing visual odometry.

ROS packages useful for computer vision tasks

The great advantage of doing computer vision in ROS is the fact that we do not have to reinvent the wheel. A lot of third-party software is available and we can also connect our vision stuff to the real robots or do some simulations. Here, we are going to enumerate some interesting computer vision tools for some of the most common visual tasks, but we will only explain one of them in detail later on (including all the steps to set it up). That is the visual odometry, but other tasks are also easy to install and start playing with. Just follow the tutorials or manuals in the links provided in the following list:

- **Visual Servoing** (also known as **Vision-based Robot Control**): This is a technique that uses feedback information obtained from a vision sensor to control the motion of a robot, typically an arm used for grasping. In ROS we have a wrapper of the **Visual Servoing Platform** (**ViSP**) software (`http://www.irisa.fr/lagadic/visp/visp.html` and `http://www.ros.org/wiki/vision_visp`). ViSP is a complete cross-platform library that allows prototyping and developing applications in visual tracking and visual servoing. The ROS wrapper provides a tracker that can be run with the node `visp_tracker` (the moving edge tracker), as well as `visp_auto_tracker` (the model-based tracker). It also helps to calibrate the camera and perform the hand-to-eye calibration, which is crucial for visual servoing in grasping tasks.
- **Augmented Reality** (**AR**): An Augmented Reality application involves the overlay of virtual imagery on the real world. A well-known library for this purpose is **ARToolkit** (`http://www.hitl.washington.edu/artoolkit/`). The main problem in this application is the tracking of the user viewpoint, so the virtual imagery is drawn in the viewpoint where the user is looking in the real world. ARToolkit video tracking libraries calculate the real camera position and orientation relative to physical markers in real time. In ROS we have a wrapper named **ar_pose** (`http://www.ros.org/wiki/ar_pose`). It allows us to track single or multiple markers at places where we can render our virtual imagery (for example, a 3D model).

- **Perception and object recognition**: The most basic perception and object recognition is possible with OpenCV libraries. However, it is worth mentioning a tool called **RoboEarth** (`http://www.roboearth.org`), which allows us to detect and build 3D models of physical objects and store them in a global database accessible to any robot (or human) worldwide. The models stored can be 2D or 3D, and can be used to recognize similar objects and their viewpoint, that is, to identify what the camera/robot is watching. The RoboEarth project is integrated in ROS, and many tutorials are provided to have a running system (`http://www.ros.org/wiki/roboearth`).
- **Visual odometry**: A visual odometry algorithm uses the images of the environment to track some features and estimate the robot movement, assuming a static environment. It can solve the 6 DoF pose of the robot with a monocular or stereo system, but it may require additional information in the monocular case. There are two main libraries for visual odometry: `libviso2` (`http://www.cvlibs.net/software/libviso2.html`) and `libfovis` (`http://www.ros.org/wiki/fovis_ros`), both of them with wrappers for ROS. The wrappers just expose these libraries to ROS. They are the stacks `viso2` and `fovis`, respectively.

  In the following section we will see how to perform visual odometry with our homemade stereo camera using the `viso2_ros` node of `viso2`. The `libviso2` library allows us to perform monocular and stereo visual odometry. However, for monocular odometry we also need the pitch and heading for the floor plane estimation. You can try the monocular case with one camera and an IMU (see *Chapter 4, Using Sensors and Actuators with ROS*), but you will always have better results with a good stereo pair, correctly calibrated, as seen thus far in this chapter. Finally, `libfovis` does not allow the monocular case, but it supports RGB-D cameras, such as the Kinect sensor (see *Chapter 4, Using Sensors and Actuators with ROS*). Regarding the stereo case, it is possible to try both libraries and see which one works better in your case. Here, we provide you with a step-by-step tutorial to install and run `viso2` in ROS.

Performing visual odometry with viso2

The current version of the `viso2` ROS wrapper builds in both ROS Fuerte and Groovy versions. However, for both of them, we must use the `catkin` building system. In this book we have seen the `rosmake` classical building system, so we provide detailed instructions for the `catkin` installation of `viso2` here. Run all the following commands in sequence:

```
cd ~
mkdir ros
```

```
cd ros
mkdir -p caktin_ws/src
cd caktin_ws/src/
catkin_init_workspace
```

Now that we have created our catkin workspace, we proceed to install viso2. We are going to do it with wstool, which integrates the downloading command in the system. This means that instead we could simply run a git clone git://github.com/srv/viso2.git, which clones the repository of the viso2 wrapper. We first install wstool and download viso2 with:

```
sudo apt-get install python-wstool
wstool init
wstool set viso2 --git git://github.com/srv/viso2.git
wstool update
```

With catkin, we must select the environment variables we want to use most frequently. We want viso2 to be installed system-wide, so we run:

```
source /opt/ros/groovy/setup.bash
cd ..
catkin_make
```

Now, with viso2 installed, we change to the developing environment variables and can run the viso2_ros nodes, such as stereo_odometer, which is the one we are going to use here. But before that, we need to publish the frame transformation between our camera and the robot or its base link. The stereo camera driver is already prepared for that, but we will explain how it is done in the following sections.

```
source devel/setup.bash
```

Camera pose calibration

In order to set the transformation between the different frames in our robot system, we must publish the tf message of such transforms. The most appropriate and generic way to do this consists of using the camera_pose stack. We use the latest version from this repository available at: https://github.com/jbohren-forks/camera_pose, because the one in ROS has some problems (for example, it fails when you run it in ROS Fuerte, since the .launch files seem to be obsolete). This stack offers a series of .launch files that calibrates the camera poses with regard to each other. It comes with .launch files for two, three, four, or more cameras. In our case we only have two cameras (stereo), so we proceed this way. First, we extend our camera_stereo.launch file with the calibrate_pose argument that calls the caliration_tf_publisher.launch file from camera_pose:

```
<include file="$(find camera_pose_calibration)/blocks/calibration_tf_
publisher.launch">
    <arg name="cache_file" value="/tmp/camera_pose_calibration_cache.
bag"/>
</include>
```

Now, run the following command:

```
roslaunch chapter6_tutorials camera_stereo.launch calibrate_pose:=true
```

And the `calibration_tf_publisher` file will publish the frame transforms (`tf`) as soon as the calibration has been done correctly. The calibration is similar to the one we have seen thus far, but using the specific tools from `camera_pose`, which are run with the following command:

```
roslaunch camera_pose_calibration calibrate_2_camera.launch camera1_ns:=/
stereo/left camera2_ns:=/stereo/right checker_rows:=6 checker_cols:=8
checker_size:=0.03
```

With this call, we can use the same calibration pattern we used with our previous calibration tools. However, it requires the images to be static; some bars move from one side to another of the image and turn green when the images in all cameras have been static for a sufficient period of time. With our noisy cameras, we need a support for the calibration pattern, a tripod or a panel, as shown in the following image:

The following image shows a tripod used simply to support the calibration pattern. This is important, since at least the camera or the calibration pattern must be fixed to a certain pose:

Then, we can calibrate as shown in the following screenshot:

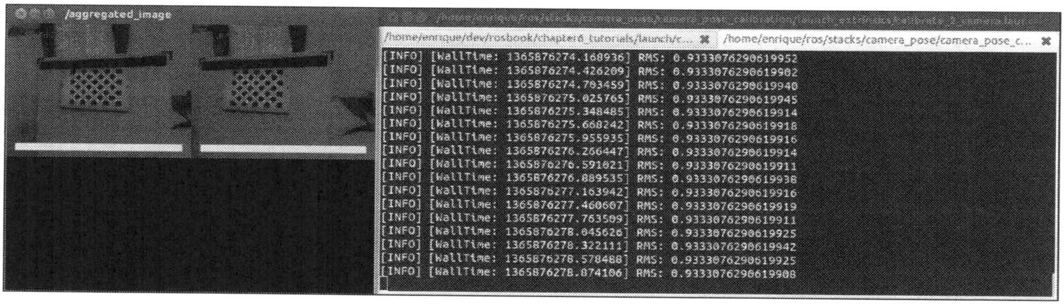

Also, this creates a `tf` frame transform from the left to the right camera. However, although this is the most appropriate way to perform the camera pose calibration, we are going to use a simple approach that is enough for a stereo pair, and is also required by `viso2`, since it just needs the frame of the whole stereo pair as a single unit/sensor. Internally, it uses the stereo calibration results of `cameracalibrator.py` to retrieve the baseline.

We have a `.launch` file that uses `static_transform_publisher` for the camera link to the base link (for example, robot base) and another one from the camera link to the optical camera link, because the optical one requires a rotation. Recall that the camera frame has the *z* axis pointing forward from the camera optical lens, while the other frames (for example, the world, navigation, or odometry) have the *z* axis pointing up. This `.launch` file is in `launch/frames/stereo_frames.launch`:

```
<launch>
  <arg name="camera" default="stereo" />

  <arg name="baseline/2" value="0.06"/>
  <arg name="optical_translation" value="0 -$(arg baseline/2) 0"/>

  <arg name="pi/2" value="1.5707963267948966"/>
  <arg name="optical_rotation" value="-$(arg pi/2) 0 -$(arg pi/2)"/>

  <node pkg="tf" type="static_transform_publisher" name="$(arg
camera)_link"
        args="0 0 0.1 0 0 0 /base_link /$(arg camera) 100"/>
  <node pkg="tf" type="static_transform_publisher" name="$(arg
camera)_optical_link"
        args="$(arg optical_translation) $(arg optical_rotation)
/$(arg camera) /$(arg camera)_optical 100"/>
</launch>
```

This file is included in our stereo camera launch file and publishes these static frame transforms. Hence, we only have to run the following command to have it publishing them:

```
roslaunch chapter6_tutorials camera_stereo.launch tf:=true
```

Then, you can check whether they are being published in `rviz` with the `tf` element, as we will see in the following section.

Running the viso2 online demo

At this point, we are reading to run the visual odometry algorithm. Our stereo pair cameras are calibrated, their frame has the appropriate name for `viso2` (ending with `_optical`), and the `tf` parameter for the camera and optical frames are published. We do not need anything else. But before using our own stereo pair, we are going to test `viso2` with the bag files provided at `ftp://opt.uib.es/bagfiles/viso2_ros`. Just run `bag/viso2_demo/download_amphoras_pool_bag_files.sh` to obtain all the bag files (it is about 4 GB). Then, we have a `.launch` file for both the monocular and stereo odometer in `launch/visual_odometry`. In order to run the stereo demo we have a `.launch` file on top that plays the bag files and also allows us to inspect and visualize its contents. For instance, to calibrate the disparity image algorithm, run the following command:

```
roslaunch chapter6_tutorials viso2_demo.launch config_disparity:=true
view:=true
```

You will see the left, right, and disparity images, and the `reconfigure_gui` interface to configure the disparity algorithm. You need to do this tuning because the bag files only have the RAW images. We have found some good parameters that are in `params/viso2_demo/disparity.yaml`. In the following screenshot you can see the results obtained with them, where you can clearly appreciate the depth of the rocks in the stereo images:

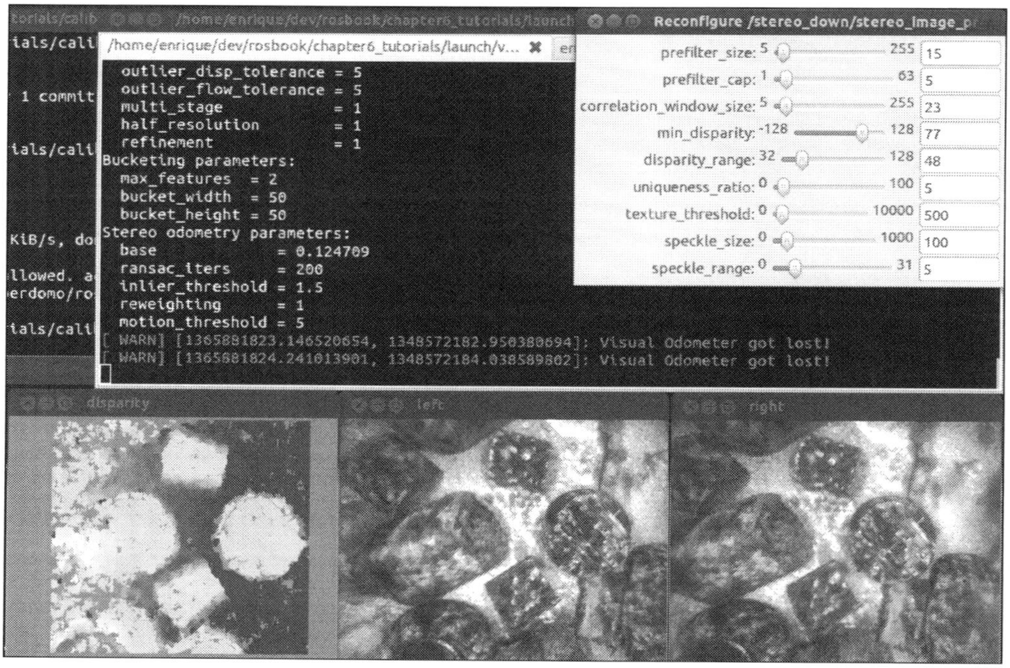

To run the stereo odometry and see the result in `rviz`, run the following command:

```
roslaunch chapter6_tutorials viso2_demo.launch odometry:=true rviz:=true
```

Note that we provide an adequate configuration for `rviz` in `params/viso2_demo/rviz.cfg`, which is automatically loaded by the launch file. The following three images show different instants of the texturized 3D point cloud (of the camera) and the `/odom` and `/stereo_optical` frames that show the camera pose estimate of the stereo odometer. The third image has a decay time of 3 seconds for the point cloud, so we can see how the points overlay over time. This way, with good images and a good odometry, we can even see a map drawn in `rviz`:

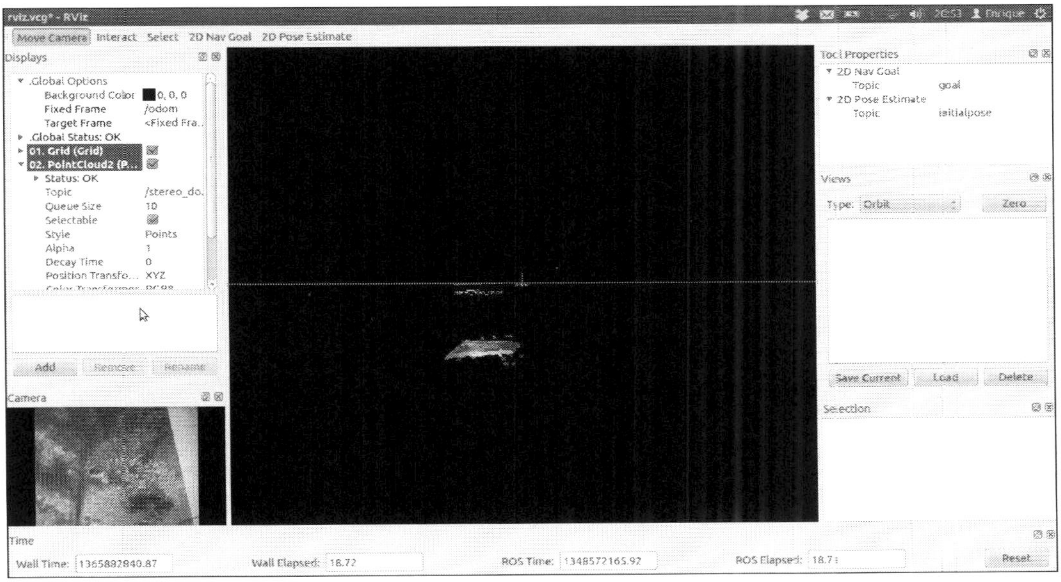

In the following screenshot we see a closer view at a later time, where we appreciate a good reconstruction, although there is some drift in the orientation (the inclination of the 3D reconstruction, which should be almost horizontal):

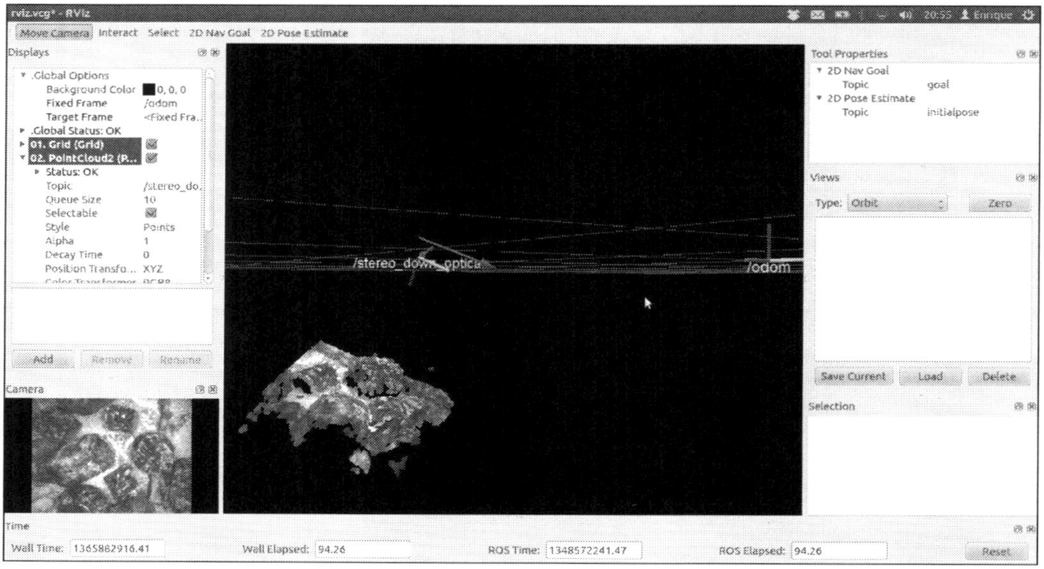

Finally, in the following image we have another view of the 3D reconstruction using the visual odometry:

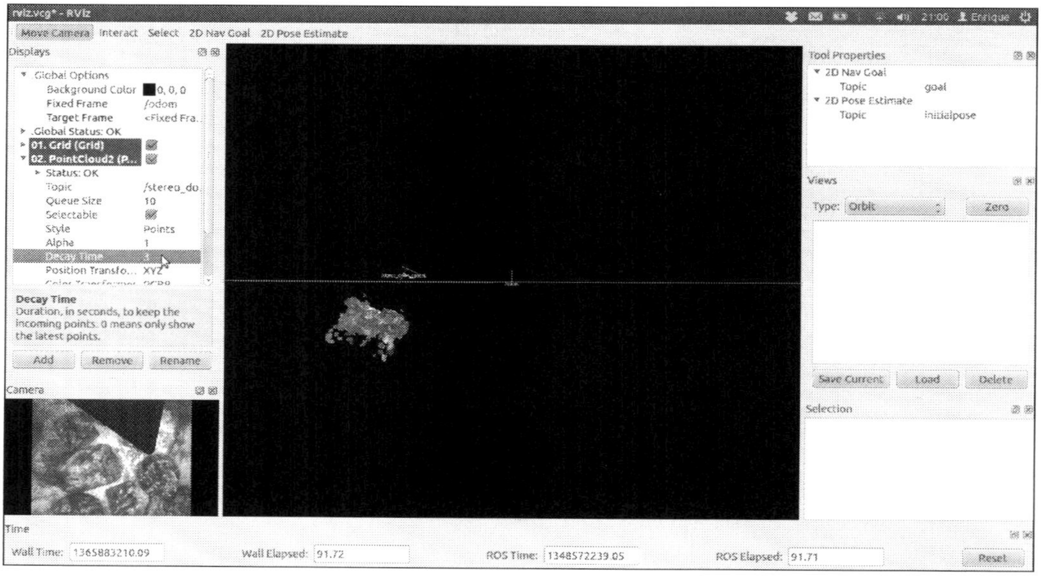

Running viso2 with our low-cost stereo camera

Finally, we can do the same as `viso2_demo` with our own stereo pair. We only have to run the following command to run the stereo odometry and see the results in `rviz` (not that the frame transforms (`tf`) are published by default):

```
roslaunch chapter6_tutorials camera_stereo.launch odometry:=true
rviz:=true
```

The following screenshot shows an example of the visual odometry system running for our low-cost stereo camera. If you move the camera, you should see the `/odom` frame moving. If the calibration is bad or the cameras are very noisy, the odometer may get lost, which is relayed to you through a warning message on the terminal. In that case, you should look for better cameras or recalibrate them to see whether better results are obtained. You will also have to look for better parameters for the disparity algorithm:

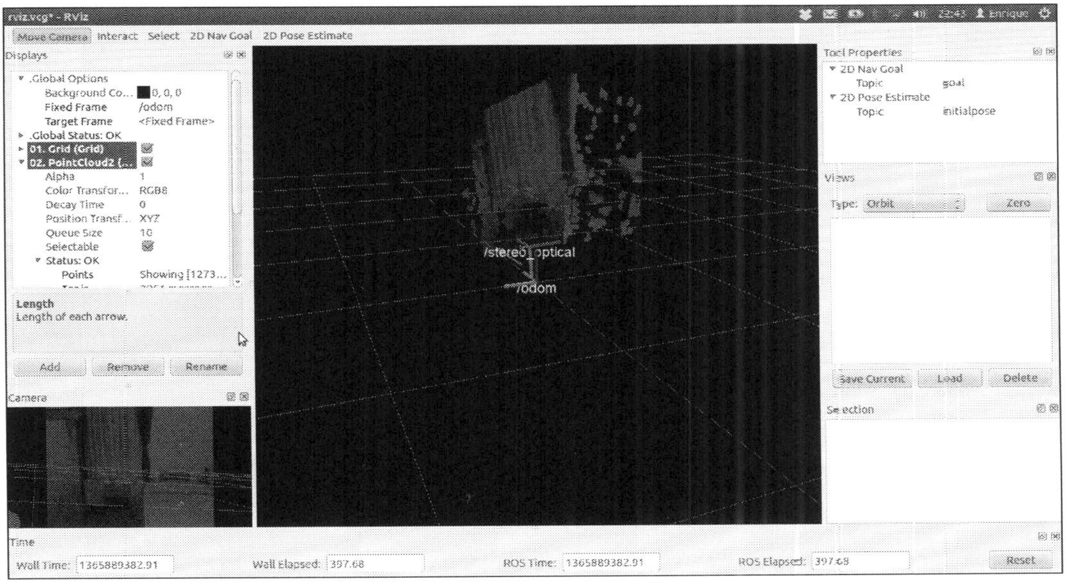

Summary

In this chapter we have given an overview of the computer vision tools provided by ROS. We started by showing how to connect and run several types of cameras, particularly FireWire and the USB ones. The basic functionality to change their parameters is presented, so now you can adjust some parameters to obtain images of good quality. Additionally, we provided a complete USB camera driver example.

Then, we moved to the camera calibration topic. With this you have learned how easy it is to calibrate a camera. The importance of calibration is the ability to correct the distortion of wide angle cameras, particularly cheap ones. Also, the calibration matrix allows you to perform many computer vision tasks, such as visual odometry or perception.

We have shown how to work with stereo vision in ROS, and how to set up an easy solution with two inexpensive webcams. We have also explained the image pipeline and several APIs that work with computer vision in ROS, such as `cv_bridge`, ImageTransport, and the integration of OpenCV within ROS packages.

Finally, we enumerated some useful tasks or topics in computer vision that are supported by some tools that are developed in ROS. In particular, we illustrated the example of visual odometry using the `viso2` library. We showed an example with some data recorded with a high quality camera and also with the inexpensive stereo pair proposed. Therefore, after reading and running the code in this chapter, you will have started with computer vision and you will now be able to perform ground-breaking stuff in minutes.

7
Navigation Stack – Robot Setups

In the previous chapters we have seen how to create our robot, mount some sensors and actuators, and move it through the virtual world using a joystick or the keyboard. Now, in this chapter, you will learn something that is probably one of the most powerful features in ROS, something that will let you move your robot autonomously.

Thanks to the community and the shared code, ROS has many algorithms that can be used for navigation.

First of all, in this chapter, you will learn all the necessary ways to configure the navigation stack with your robot. In the next chapter, you will learn to configure and launch the navigation stack on the simulated robot, giving goals and configuring some parameters to get the best results. In particular, we will cover the following items in this chapter:

- Introduction to the navigation stacks and their powerful capabilities—clearly one of the greatest pieces of software that comes with ROS.
- The TF is explained in order to show how to transform from the frame of one physical element to the other; for example, the data received using a sensor or the command for the desired position of an actuator.
- We will see how to create a laser driver or simulate it.
- We will learn how the odometry is computed and published, and how Gazebo provides it.
- A base controller will be presented, including a detailed description of how to create one for your robot.

- We will see how to execute SLAM with ROS. That is, we will show you how you can build a map from the environment with your robot as it moves through it.
- Finally, you will be able to localize your robot in the map using the localization algorithms of the navigation stack.

The navigation stack in ROS

In order to understand the navigation stack, you should think of it as a set of algorithms that use the sensors of the robot and the odometry, and you can control the robot using a standard message. It can move your robot without problems (for example, without crashing or getting stuck in some location, or getting lost) to another position.

You would assume that this stack can be easily used with any robot. This is almost true, but it is necessary to tune some configuration files and write some nodes to use the stack.

The robot must satisfy some requirements before it uses the navigation stack:

- The navigation stack can only handle a differential drive and holonomic-wheeled robots. The shape of the robot must be either a square or a rectangle. However, it can also do certain things with biped robots, such as robot localization, as long as the robot does not move sideways.
- It requires that the robot publishes information about the relationships between all the joints and sensors' position.
- The robot must send messages with linear and angular velocities.
- A planar laser must be on the robot to create the map and localization. Alternatively, you can generate something equivalent to several lasers or a sonar, or you can project the values to the ground if they are mounted in another place on the robot.

The following diagram shows you how the navigation stacks are organized. You can see three groups of boxes with colors (gray and white) and dotted lines. The plain white boxes indicate those stacks that are provided by ROS, and they have all the nodes to make your robot really autonomous:

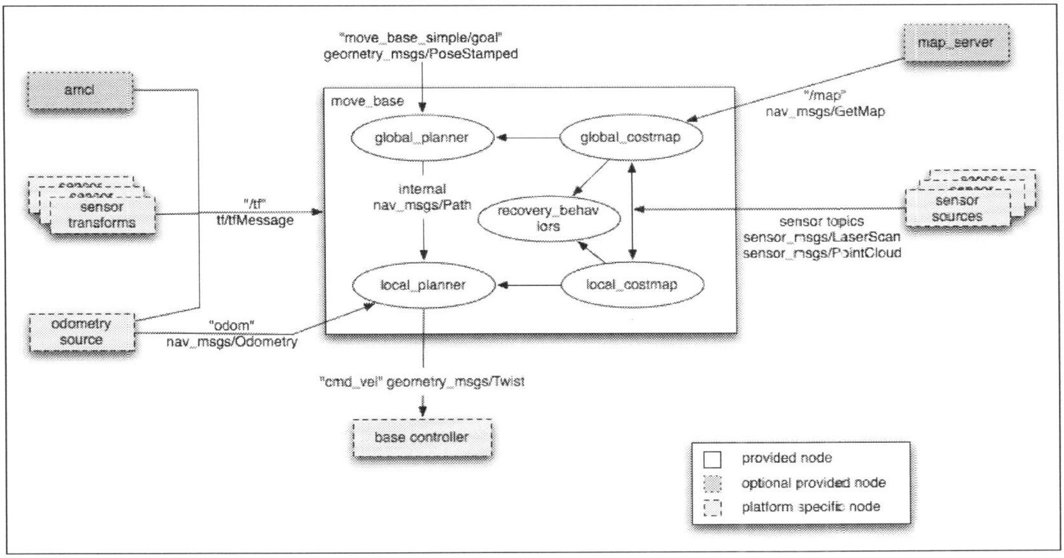

In the following sections, we will see how to create the parts marked in gray in the diagram. These parts depend on the platform used; this means that it is necessary to write code to adapt the platform to be used in ROS and to be used by the navigation stack.

Creating transforms

The navigation stack needs to know the position of the sensors, wheels, and joints.

To do that, we use the **TF** (which stands for **Transform Frames**) software library. It manages a transform tree. You could do this with mathematics, but if you have a lot of frames to calculate, it will be a bit complicated and messy.

Thanks to TF, we can add more sensors and parts to the robot, and the TF will handle all the relations for us.

If we put the laser 10 cm backwards and 20 cm above with regard to the origin of the coordinates of `base_link`, we would need to add a new frame to the transformation tree with these offsets.

Once inserted and created, we could easily know the position of the laser with regard to the `base_link` value or the wheels. The only thing we need to do is call the `TF` library and get the transformation.

Creating a broadcaster

Let's test it with a simple code. Create a new file in `chapter7_tutorials/src` with the name `tf_broadcaster.cpp`, and put the following code inside it:

```
#include <ros/ros.h>
#include <tf/transform_broadcaster.h>

int main(int argc, char** argv){
  ros::init(argc, argv, "robot_tf_publisher");
  ros::NodeHandle n;

  ros::Rate r(100);

  tf::TransformBroadcaster broadcaster;

  while(n.ok()){
    broadcaster.sendTransform(
      tf::StampedTransform(
        tf::Transform(tf::Quaternion(0, 0, 0, 1), tf::Vector3(0.1, 
0.0, 0.2)),
        ros::Time::now(),"base_link", "base_laser"));
    r.sleep();
  }
}
```

Remember to add the following line in your `CMakelist.txt` file to create the new executable:

```
rosbuild_add_executable(tf_broadcaster src/tf_broadcaster.cpp)
```

And we also create another node that will use the transform, and it will give us the position of a point of a sensor with regard to the center of `base_link` (our robot).

Creating a listener

Create a new file in `chapter7_tutorials/src` with the name `tf_listener.cpp` and input the following code:

```
#include <ros/ros.h>
#include <geometry_msgs/PointStamped.h>
#include <tf/transform_listener.h>

void transformPoint(const tf::TransformListener& listener){
  //we'll create a point in the base_laser frame that we'd like to 
transform to the base_link frame
  geometry_msgs::PointStamped laser_point;
  laser_point.header.frame_id = "base_laser";
```

```
  //we'll just use the most recent transform available for our simple
example
  laser_point.header.stamp = ros::Time();

  //just an arbitrary point in space
  laser_point.point.x = 1.0;
  laser_point.point.y = 2.0;
  laser_point.point.z = 0.0;

    geometry_msgs::PointStamped base_point;
    listener.transformPoint("base_link", laser_point, base_point);

    ROS_INFO("base_laser: (%.2f, %.2f. %.2f) -----> base_link: (%.2f,
%.2f, %.2f) at time %.2f",
        laser_point.point.x, laser_point.point.y, laser_point.point.z,
        base_point.point.x, base_point.point.y, base_pcint.point.z,
base_point.header.stamp.toSec());

    ROS_ERROR("Received an exception trying to transform a point from
\"base_laser\" to \"base_link\": %s", ex.what());

}

int main(int argc, char** argv){
  ros::init(argc, argv, "robot_tf_listener");
  ros::NodeHandle n;

  tf::TransformListener listener(ros::Duration(10));

  //we'll transform a point once every second
  ros::Timer timer = n.createTimer(ros::Duration(1.0),
boost::bind(&transformPoint, boost::ref(listener)));

  ros::spin();

}
```

Remember to add the line in the `CMakeList.txt` file to create the executable.

Compile the package and run both the nodes using the following commands:

```
$ rosmake chapter7_tutorials
$ rosrun chapter7_tutorials tf_broadcaster
$ rosrun chapter7_tutorials tf_listener
```

Then you will see the following message:

```
[ INFO] [1368521854.336910465]: base_laser: (1.00, 2.00. 0.00) ----->
base_link: (1.10, 2.00, 0.20) at time 1368521854.33
```

```
[ INFO] [1368521855.336347545]: base_laser: (1.00, 2.00. 0.00) ----->
base_link: (1.10, 2.00, 0.20) at time 1368521855.33
```

This means that the point that you published on the node, with the position (1.00, 2.00, 0.00) relative to `base_laser`, has the position (1.10, 2.00, 0.20) relative to `base_link`.

As you can see, the `tf` library performs all the mathematics for you to get the coordinates of a point or the position of a joint relative to another point.

A transform tree defines offsets in terms of both translation and rotation between different coordinate frames.

Let us see an example to help you understand this. In our robot model used in *Chapter 5, 3D Modeling and Simulation*, we are going to add another laser, say, on the back of the robot (`base_link`):

The system had to know the position of the new laser to detect collisions, such as the one between wheels and walls. With the TF tree, this is very simple to do and maintain and is also scalable. Thanks to `tf`, we can add more sensors and parts, and the `tf` library will handle all the relations for us. All the sensors and joints must be correctly configured on `tf` to permit the navigation stack to move the robot without problems, and to exactly know where each one of their components is.

Before starting to write the code to configure each component, keep in mind that you have the geometry of the robot specified in the URDF file. So, for this reason, it is not necessary to configure the robot again. Perhaps you do not know it, but you have been using the `robot_state_publisher` package to publish the transform tree of your robot. In *Chapter 5, 3D Modeling and Simulation*, we used it for the first time; therefore, you do have the robot configured to be used with the navigation stack.

Watching the transformation tree

If you want to see the transformation tree of your robot, use the following command:

```
$ roslaunch chapter7_tutorials gazebo_map_robot.launch model:="`rospack find chapter7_tutorials`/urdf/robot1_base_04.xacro"
$ rosrun tf view_frames
```

The resultant frame is depicted as follows:

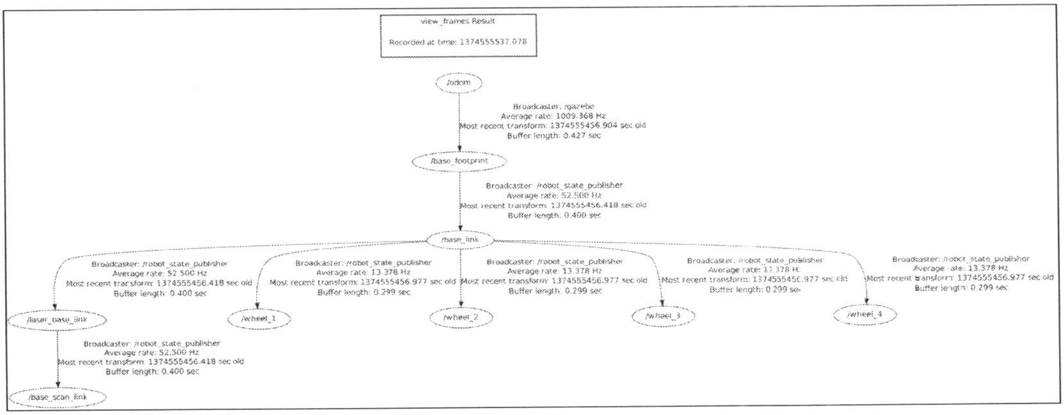

And now, if you run `tf_broadcaster` and run the `rosrun tf view_frames` command again, you will see the frame that you have created by code:

```
$ rosrun chapter7_tutorials tf_broadcaster
$ rosrun tf view_frames
```

> **Downloading the example code**
>
> You can download the example code files for all Packt books you have purchased from your account at `http://www.packtpub.com`. If you purchased this book elsewhere, you can visit `http://www.packtpub.com/support` and register to have the files e-mailed directly to you. You can also download these code files from `https://github.com/AaronMR/Learning_ROS_for_Robotics_Programming`.

The resultant frame is depicted as follows:

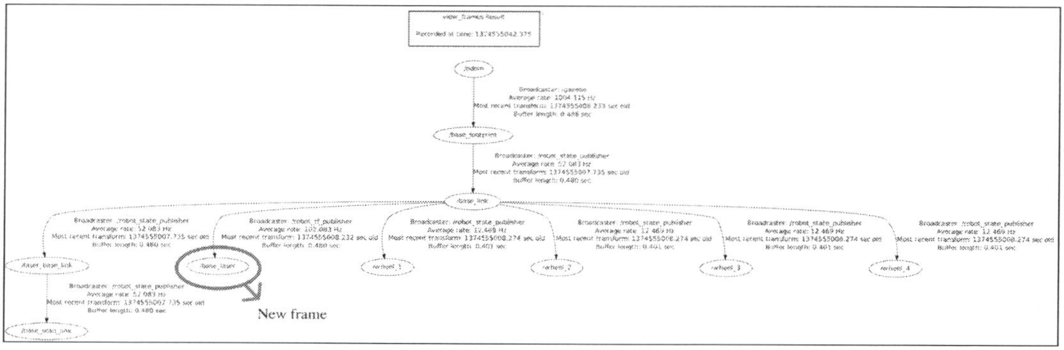

Publishing sensor information

Your robot can have a lot of sensors to see the world; you can program a lot of nodes to take these data and do something, but the navigation stack is prepared only to use the planar laser's sensor. So, your sensor must publish the data with one of these types: `sensor_msgs/LaserScan` or `sensor_msgs/PointCloud`.

We are going to use the laser located in front of the robot to navigate in Gazebo. Remember that this laser is simulated on Gazebo, and it publishes data on the `base_scan/scan` frame.

In our case, we do not need to configure anything of our laser to use it on the navigation stack. This is because we have `tf` configured in the `.urdf` file, and the laser is publishing data with the correct type.

If you use a real laser, ROS might have a driver for it. Indeed, in *Chapter 4, Using Sensors and Actuators with ROS*, you learned how to connect the Hokuyo laser to ROS. Anyway, if you are using a laser that has no driver on ROS and want to write a node to publish the data with the `sensor_msgs/LaserScan` sensor, you have an example template to do it, which is shown in the following section.

But first, remember the structure of the message `sensor_msgs/LaserScan`. Use the following command:

```
$ rosmsg show sensor_msgs/LaserScan

std_msgs/Header header
  uint32 seq
  time stamp
```

```
  string frame_id
float32 angle_min
float32 angle_max
float32 angle_increment
float32 time_increment
float32 scan_time
float32 range_min
float32 range_max
float32[] ranges
float32[] intensities
```

Creating the laser node

Now we will create a new file in `chapter7_tutorials/src` with the name `laser.cpp` and put the following code in it:

```
#include <ros/ros.h>
#include <sensor_msgs/LaserScan.h>

int main(int argc, char** argv){
  ros::init(argc, argv, "laser_scan_publisher");

  ros::NodeHandle n;
  ros::Publisher scan_pub = n.advertise<sensor_
msgs::LaserScan>("scan", 50);

  unsigned int num_readings = 100;
  double laser_frequency = 40;
  double ranges[num_readings];
  double intensities[num_readings];

  int count = 0;
  ros::Rate r(1.0);
  while(n.ok()){
    //generate some fake data for our laser scan
    for(unsigned int i = 0; i < num_readings; ++i){
      ranges[i] = count;
      intensities[i] = 100 + count;
    }
    ros::Time scan_time = ros::Time::now();
```

```
        //populate the LaserScan message
        sensor_msgs::LaserScan scan;
        scan.header.stamp = scan_time;
        scan.header.frame_id = "base_link";
        scan.angle_min = -1.57;
        scan.angle_max = 1.57;
        scan.angle_increment = 3.14 / num_readings;
        scan.time_increment = (1 / laser_frequency) / (num_readings);
        scan.range_min = 0.0;
        scan.range_max = 100.0;

        scan.ranges.resize(num_readings);
        scan.intensities.resize(num_readings);
        for(unsigned int i = 0; i < num_readings; ++i){
          scan.ranges[i] = ranges[i];
          scan.intensities[i] = intensities[i];
        }

        scan_pub.publish(scan);
        ++count;
        r.sleep();
      }
    }
```

As you can see, we are going to create a new topic with the name `scan` and the message type `sensor_msgs/LaserScan`. You must be familiar with this message type from *Chapter 4, Using Sensors and Actuators with ROS*. The name of the topic must be unique. When you configure the navigation stack, you will select this topic to be used for the navigation. The following command line shows how to create the topic with the correct name:

```
ros::Publisher scan_pub = n.advertise<sensor_msgs::LaserScan>("scan",
50);
```

It is important to publish data with `header`, `stamp`, `frame_id`, and many more elements because, if not, the navigation stack could fail with such data:

```
scan.header.stamp = scan_time;
scan.header.frame_id = "base_link";
```

Other important data on `header` is `frame_id`. It must be one of the frames created in the `.urdf` file and must have a frame published on the `tf` frame transforms. The navigation stack will use this information to know the real position of the sensor and make transforms such as the one between the data sensor and obstacles.

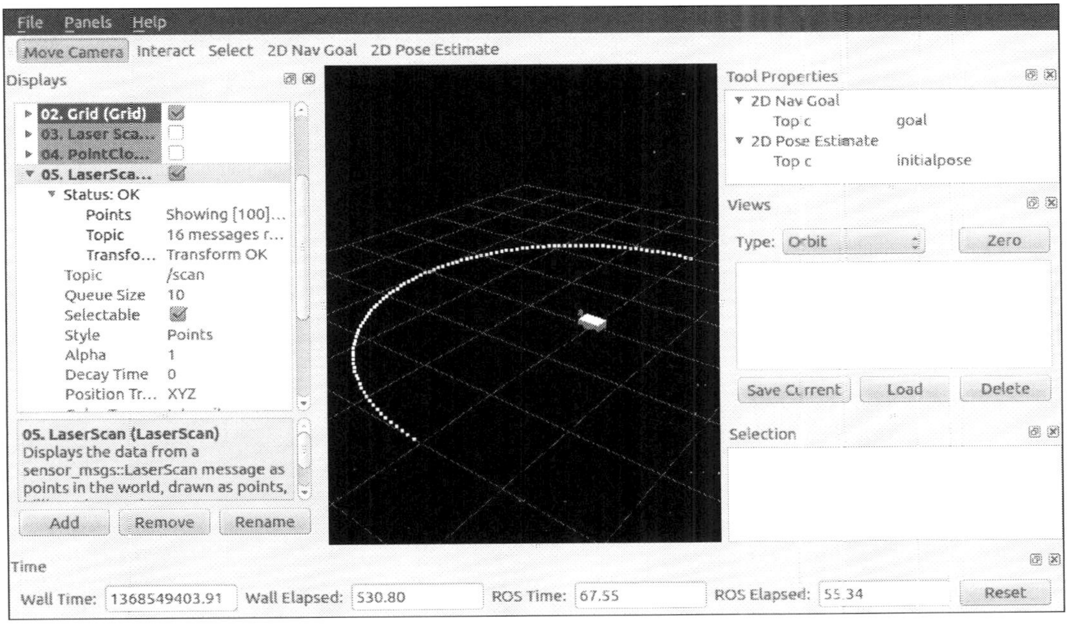

With this template, you can use any laser although it has no driver for ROS. You only have to change the fake data with the right data from your laser.

This template can also be used to create something that looks like a laser but is not. For example, you could simulate a laser using stereoscopy or using a sensor such as a sonar.

Publishing odometry information

The navigation stack also needs to receive data from the robot odometry. The odometry is the distance of something relative to a point. In our case, it is the distance between `base_link` and a fixed point in the frame `odom`.

The type of message used by the navigation stack is `nav_msgs/Odometry`. We are going to watch the structure using the following command:

```
$ rosmsg show nav_msgs/Odometry
```

```
std_msgs/Header header
  uint32 seq
  time stamp
  string frame_id
string child_frame_id
geometry_msgs/PoseWithCovariance pose
  geometry_msgs/Pose pose
    geometry_msgs/Point position
      float64 x
      float64 y
      float64 z
    geometry_msgs/Quaternion orientation
      float64 x
      float64 y
      float64 z
      float64 w
  float64[36] covariance
geometry_msgs/TwistWithCovariance twist
  geometry_msgs/Twist twist
    geometry_msgs/Vector3 linear
      float64 x
      float64 y
      float64 z
    geometry_msgs/Vector3 angular
      float64 x
      float64 y
      float64 z
  float64[36] covariance
```

As you can see in the message structure, `nav_msgs/Odometry` gives the position of the robot between `frame_id` and `child_frame_id`. It also gives us the pose of the robot using the `geometry_msgs/Pose` message and the velocity with the `geometry_msgs/Twist` message.

The pose has two structures that show the position in Euler coordinates and the orientation of the robot using a quaternion. The orientation is the angular displacement of the robot.

The velocity has two structures that show the linear velocity and the angular velocity. For our robot, we will use only the linear *x* velocity and the angular *z* velocity. We will use the linear *x* velocity to know whether the robot is moving forward or backward. The angular *z* velocity is used to check whether the robot is rotating towards the left or right.

As the odometry is the displacement between two frames, it is necessary to publish the transform of it. We did it in the last point, but later on in this section, I will show you an example to publish the odometry and `tf` of our robot.

Now let me show you how Gazebo works with the odometry.

How Gazebo creates the odometry

As you have seen in other examples with Gazebo, our robot moves in the simulated world just like a robot in the real world. We use a driver for our robot, the `diffdrive_plugin`. We configured this plugin in *Chapter 5, 3D Modeling and Simulation*, when you created the robot to use it in Gazebo.

This driver publishes the odometry generated in the simulated world, so we do not need to write anything for Gazebo.

Execute the robot sample in Gazebo to see the odometry working. Type the following commands in the shell:

```
$ roslaunch chapter7_tutorials gazebo_xacro.launch model:="`rospack find chapter7_tutorials`/urdf/robot1_base_04.xacro"
$ rosrun erratic_teleop erratic_keyboard_teleop
```

Then, with the teleop node, move the robot for a few seconds to generate new data on the odometry topic.

On the screen of the Gazebo simulator, if you click on **robot_model1**, you will see some properties of the object. One of these properties is the pose of the robot. Click on the pose, and you will see some fields with data. What you are watching is the position of the robot in the virtual world. If you move the robot, the data changes:

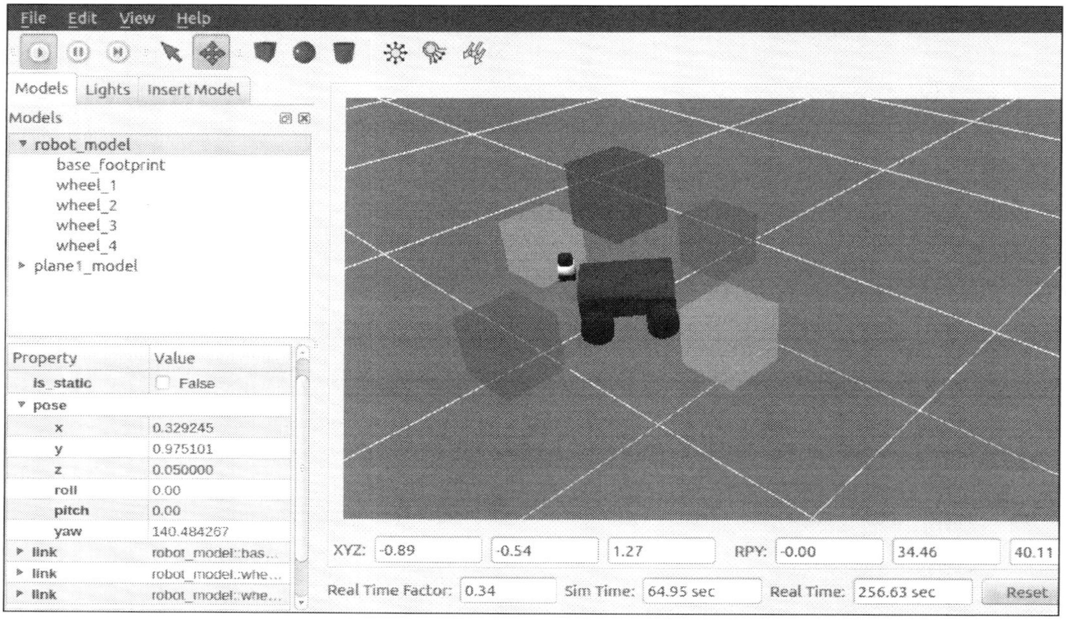

Gazebo continuously publishes the odometry data. Check the topic and see what data it is sending. Type the following command in a shell:

```
$ rostopic echo /odom/pose/pose
```

The following is the output you will receive:

```
---
position:
  x: 0.32924763712
  y: 0.97509878254
  z: 0.0
orientation:
  x: 0.0
  y: 0.0
  z: 0.941128847661
  w: 0.33804806182
---
```

> Notice that the data is the same as the one you can see on the Gazebo screen.

As you can observe, Gazebo is creating the odometry as the robot moves. We are going to see how Gazebo creates it by looking inside the plugin's source code.

The plugin file is located in the `erratic_gazebo_plugins` package, and the file is `diffdrive_plugin.cpp`. Open the file and you will see the following code inside the file:

```
$ rosed erratic_gazebo_plugins diffdrive_plugin.cpp
```

The file has a lot of code, but the important part for us now is the following function, `publish_odometry()`:

```
void DiffDrivePlugin::publish_odometry()
{
  ros::Time current_time = ros::Time::now();
  std::string odom_frame = tf::resolve(tf_prefix_, "odom");
  std::string base_footprint_frame = tf::resolve(tf_prefix_, "base_
footprint");

  // getting data for base_footprint to odom transform
  math::Pose pose = this->parent->GetState().GetPose();

  btQuaternion qt(pose.rot.x, pose.rot.y, pose.rot.z, pose.rot.w);
  btVector3 vt(pose.pos.x, pose.pos.y, pose.pos.z);

  tf::Transform base_footprint_to_odom(qt, vt);
  transform_broadcaster_->sendTransform(tf::StampedTransform(base_
footprint_to_odom, current_time, odom_frame, base_footprint_frame));

  // publish odom topic
  odom_.pose.pose.position.x = pose.pos.x;
  odom_.pose.pose.position.y = pose.pos.y;

  odom_.pose.pose.orientation.x = pose.rot.x;
  odom_.pose.pose.orientation.y = pose.rot.y;
  odom_.pose.pose.orientation.z = pose.rot.z;
  odom_.pose.pose.orientation.w = pose.rot.w;

  math::Vector3 linear = this->parent->GetWorldLinearVel();
  odom_.twist.twist.linear.x = linear.x;
  odom_.twist.twist.linear.y = linear.y;
  odom_.twist.twist.angular.z = this->parent->GetWorldAngularVel().z;
```

```
    odom_.header.stamp = current_time;
    odom_.header.frame_id = odom_frame;
    odom_.child_frame_id = base_footprint_frame;

    pub_.publish(odom_);
  }
```

The `publish_odometry()` function is where the odometry is published. You can see how the fields of the structure are filled and the name of the topic for the odometry is set (in this case, it is `odom`). The pose is generated in the other part of the code that we will see in the following section.

Once you have learned how and where Gazebo creates the odometry, you will be ready to learn how to publish the odometry and `tf` for a real robot. The following code will show a robot doing circles continuously. The finality does not really matter; the important thing to know is how to publish the correct data for our robot.

Creating our own odometry

Create a new file in `chapter7_tutorials/src` with the name `odometry.cpp` and put the following code in it:

```
#include <string>
#include <ros/ros.h>
#include <sensor_msgs/JointState.h>
#include <tf/transform_broadcaster.h>
#include <nav_msgs/Odometry.h>


int main(int argc, char** argv) {

ros::init(argc, argv, "state_publisher");
    ros::NodeHandle n;
    ros::Publisher odom_pub = n.advertise<nav_msgs::Odometry>("odom",
10);

    // initial position
    double x = 0.0;
    double y = 0.0;
    double th = 0;

    // velocity
    double vx = 0.4;
    double vy = 0.0;
    double vth = 0.4;
```

```
    ros::Time current_time;
    ros::Time last_time;
    current_time = ros::Time::now();
    last_time = ros::Time::now();

    tf::TransformBroadcaster broadcaster;
    ros::Rate loop_rate(20);

    const double degree = M_PI/180;

    // message declarations
    geometry_msgs::TransformStamped odom_trans;
    odom_trans.header.frame_id = "odom";
    odom_trans.child_frame_id = "base_footprint";

    while (ros::ok()) {
        current_time = ros::Time::now();

        double dt = (current_time - last_time).toSec();
        double delta_x = (vx * cos(th) - vy * sin(th)) * dt;
        double delta_y = (vx * sin(th) + vy * cos(th)) * dt;
        double delta_th = vth * dt;

        x += delta_x;
        y += delta_y;
        th += delta_th;

        geometry_msgs::Quaternion odom_quat;
        odom_quat = tf::createQuaternionMsgFromRollPitchYaw(0,0,th);

        // update transform
        odom_trans.header.stamp = current_time;
        odom_trans.transform.translation.x = x;
        odom_trans.transform.translation.y = y;
        odom_trans.transform.translation.z = 0.0;
        odom_trans.transform.rotation = tf::createQuaternionMsgFromYa
w(th);

        //filling the odometry
        nav_msgs::Odometry odom;
        odom.header.stamp = current_time;
        odom.header.frame_id = "odom";
        odom.child_frame_id = "base_footprint";
```

```
            // position
            odom.pose.pose.position.x = x;
            odom.pose.pose.position.y = y;
            odom.pose.pose.position.z = 0.0;
            odom.pose.pose.orientation = odom_quat;

            // velocity
            odom.twist.twist.linear.x = vx;
            odom.twist.twist.linear.y = vy;
            odom.twist.twist.linear.z = 0.0;
            odom.twist.twist.angular.x = 0.0;
            odom.twist.twist.angular.y = 0.0;
            odom.twist.twist.angular.z = vth;

          last_time = current_time;

            // publishing the odometry and the new tf
            broadcaster.sendTransform(odom_trans);
            odom_pub.publish(odom);

          loop_rate.sleep();
        }
        return 0;
    }
```

First, create the transformation variable and fill it with `frame_id` and the `child_frame_id` values to know when the frames have to move. In our case, the base `base_footprint` will move relatively toward the frame `odom`:

```
geometry_msgs::TransformStamped odom_trans;
    odom_trans.header.frame_id = "odom";
    odom_trans.child_frame_id = "base_footprint";
```

In this part, we generate the pose of the robot. With the linear velocity and the angular velocity, we can calculate the theoretical position of the robot after a while:

```
double dt = (current_time - last_time).toSec();
double delta_x = (vx * cos(th) - vy * sin(th)) * dt;
double delta_y = (vx * sin(th) + vy * cos(th)) * dt;
double delta_th = vth * dt;

x += delta_x;
y += delta_y;
th += delta_th;

geometry_msgs::Quaternion odom_quat;
odom_quat = tf::createQuaternionMsgFromRollPitchYaw(0,0,th);
```

> In this book, you will not find an explanation about the kinematics of the robot. You can find a lot of literature on the Internet about it; you should look out for "differential wheel kinematics".

On the transformation, we will only fill in the x and rotation fields, as our robot can only move forward and backward and can turn:

```
        odom_trans.header.stamp = current_time;
        odom_trans.transform.translation.x = x;
        odom_trans.transform.translation.y = 0.0;
        odom_trans.transform.translation.z = 0.0;
        odom_trans.transform.rotation = tf::createQuaternionMsgFromYa
w(th);
```

With the odometry, we will do the same. Fill the frame_id and child_frame_id fields with odom and base_footprint.

As the odometry has two structures, we will fill in the x, y, and orientation of the pose. On the twist structure, we will fill in the linear velocity x and the angular velocity z:

```
// position
odom.pose.pose.position.x = x;
odom.pose.pose.position.y = y;
odom.pose.pose.orientation = odom_quat;

// velocity
odom.twist.twist.linear.x = vx;
odom.twist.twist.angular.z = vth;
```

Once all the necessary fields are filled in, publish the data:

```
// publishing the odometry and the new tf
broadcaster.sendTransform(odom_trans);
odom_pub.publish(odom);
```

Remember to create the following line in the CMakeLists.txt file before you run rosmake chapter7_tutorials:

```
rosbuild_add_executable(odometry src/odometry.cpp)
```

Compile the package and launch the robot without using Gazebo, using only rviz to visualize the model and the movement of the robot. Use the following command to do this:

```
$ roslaunch chapter7_tutorials display_xacro.launch model:="`rospack find chapter7_tutorials`/urdf/robot1_base_04.xacro"
```

And run the odometry node with the following command:

```
$ rosrun chapter7_tutorials odometry
```

This is what you will get:

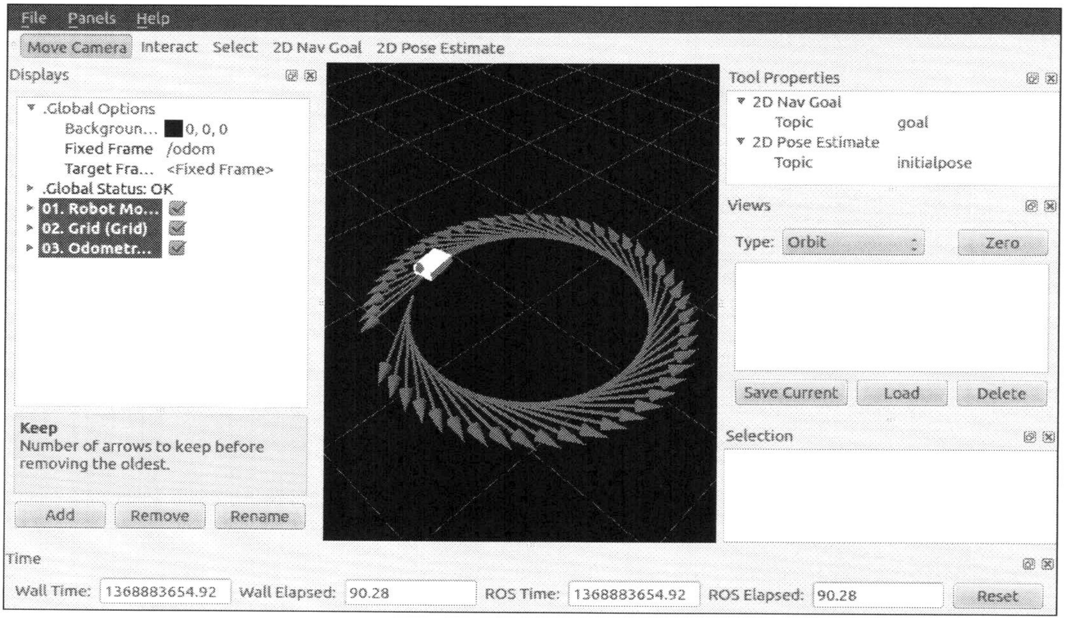

On the `rviz` screen, you can see the robot moving over some red arrows (grid). The robot moves over the grid because you published a new `tf` frame transform for the robot. The red arrows are the graphical representation for the odometry message. You will see the robot moving in circles continuously as we programmed in the code.

Creating a base controller

A base controller is an important element in the navigation stack because it is the only way to effectively control your robot. It communicates directly with the electronics of your robot.

ROS does not provide a standard base controller, so you must write a base controller for your mobile platform.

Your robot has to be controlled with the message type `geometry_msgs/Twist`. This message is used on the `Odometry` message that we have seen before.

So, your base controller must subscribe to a topic with the name `cmd_vel` and must generate the correct commands to move the platform with the correct linear and angular velocities.

We are now going to recall the structure of this message. Type the following command in a shell to see the structure:

```
$ rosmsg show geometry_msgs/Twist
```

The output of this command is as follows:

```
geometry_msgs/Vector3 linear
  float64 x
  float64 y
  float64 z
geometry_msgs/Vector3 angular
  float64 x
  float64 y
  float64 z
```

The vector with the name `linear` indicates the linear velocity for the axes `x`, `y`, and `z`. The vector with the name `angular` is for the angular velocity on the axes.

For our robot, we will only use the linear velocity `x` and the angular velocity `z`. This is because our robot is on a differential wheeled platform, and it has two motors to move the robot forward and backward and to turn.

We are working with a simulated robot on Gazebo, and the base controller is implemented on the driver used to move/simulate the platform. This means that we will not have to create the base controller for this robot.

Anyway, in this chapter, you will see an example to implement the base controller on your physical robot. Before that, let's go to execute our robot on Gazebo to see how the base controller works. Run the following commands on different shells:

```
$ roslaunch chapter7_tutorials gazebo_xacro.launch model:="`rospack find chapter7_tutorials`/urdf/robot1_base_04.xacro"
$ rosrun erratic_teleop erratic_keyboard_teleop
```

When all the nodes are launched and working, open `rxgraph` to see the relation between all the nodes:

```
$ rxgraph
```

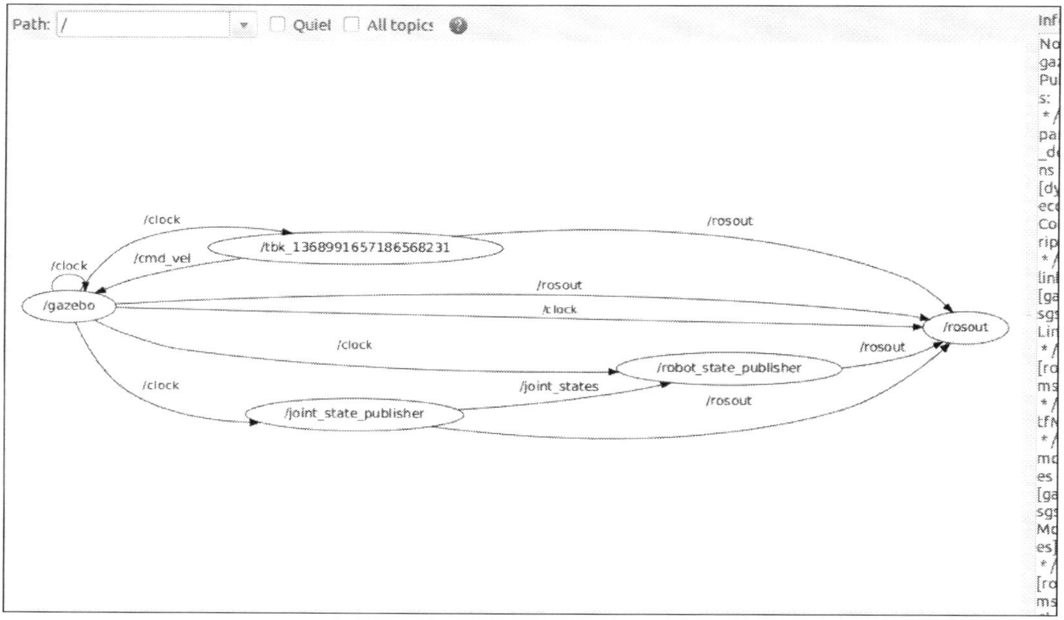

You can see that Gazebo subscribes automatically to the `cmd_vel` topic that is generated by the teleoperation node.

Inside the Gazebo simulator, the plugin of our differential wheeled robot is running and is getting the data from the `cmd_vel` topic. Also, this plugin moves the robot in the virtual world and generates the odometry.

Using Gazebo to create the odometry

To obtain some insight of how Gazebo does that, we are going to have a sneak peek inside the `diffdrive_plugin.cpp` file:

```
$ rosed erratic_gazebo_plugins diffdrive_plugin.cpp
```

The `Load(...)` function performs the subscription to the topic, and when a `cmd_vel` topic is received, the `cmdVelCallback()` function is executed to handle the message:

```
void DiffDrivePlugin::Load(physics::ModelPtr _parent, sdf::ElementPtr
_sdf)
{
...
...
ros::SubscribeOptions so =
    ros::SubscribeOptions::create<geometry_msgs::Twist>(topicName,
1, boost::bind(&DiffDrivePlugin::cmdVelCallback, this, _1),
ros::VoidPtr(), &queue_);

}
```

When a message arrives, the linear and angular velocities are stored in the internal variables to run some operations later:

```
void DiffDrivePlugin::cmdVelCallback(const geometry_
msgs::Twist::ConstPtr& cmd_msg)
{
  ...
  ...
    x_ = cmd_msg->linear.x;
    rot_ = cmd_msg->angular.z;
  ...
  ...
}
```

The plugin estimates the velocity for each motor using the formulas from the kinematic model of the robot in the following manner:

```
void DiffDrivePlugin::GetPositionCmd()
{
  ...
    vr = x_;
    va = rot_;

    wheelSpeed[LEFT] = vr + va * wheelSeparation / 2.0;
    wheelSpeed[RIGHT] = vr - va * wheelSeparation / 2.0;
  ...
}
```

And finally, it estimates the distance traversed by the robot using more formulas from the kinematic motion model of the robot. As you can see in the code, you must know the wheel diameter and the wheel separation of your robot:

```
// Update the controller
void DiffDrivePlugin::UpdateChild()
{
    ...
  ...
    wd = wheelDiameter;
    ws = wheelSeparation;

    // Distance travelled by front wheels
    d1 = stepTime * wd / 2 * joints[LEFT]->GetVelocity(0);
    d2 = stepTime * wd / 2 * joints[RIGHT]->GetVelocity(0);

    dr = (d1 + d2) / 2;
    da = (d1 - d2) / ws;

    // Compute odometric pose
    odomPose[0] += dr * cos(odomPose[2]);
    odomPose[1] += dr * sin(odomPose[2]);
    odomPose[2] += da;

    // Compute odometric instantaneous velocity
    odomVel[0] = dr / stepTime;
    odomVel[1] = 0.0;
    odomVel[2] = da / stepTime;
...
  ...
}
```

This is the way `diffdrive_plugin` controls our simulated robot in Gazebo.

Creating our base controller

Now we are going to do something similar, that is, prepare a code to be used with a real robot with two wheels and encoders.

Create a new file in `chapter7_tutorials/src` with the name `base_controller.cpp` and put in the following code:

```
#include <ros/ros.h>
#include <sensor_msgs/JointState.h>
#include <tf/transform_broadcaster.h>
```

```
#include <nav_msgs/Odometry.h>
#include <iostream>

using namespace std;

double width_robot = 0.1;
double vl = 0.0;
double vr = 0.0;
ros::Time last_time;
double right_enc = 0.0;
double left_enc = 0.0;
double right_enc_old = 0.0;
double left_enc_old = 0.0;
double distance_left = 0.0;
double distance_right = 0.0;
double ticks_per_meter = 100;
double x = 0.0;
double y = 0.0;
double th = 0.0;
geometry_msgs::Quaternion odom_quat;

void cmd_velCallback(const geometry_msgs::Twist &twist_aux)
{
    geometry_msgs::Twist twist = twist_aux;
    double vel_x = twist_aux.linear.x;
    double vel_th = twist_aux.angular.z;
    double right_vel = 0.0;
    double left_vel = 0.0;

    if(vel_x == 0){
// turning
        right_vel = vel_th * width_robot / 2.0;
        left_vel = (-1) * right_vel;
    }else if(vel_th == 0){
// forward / backward
        left_vel = right_vel = vel_x;
    }else{
// moving doing arcs
        left_vel = vel_x - vel_th * width_robot / 2.0;
        right_vel = vel_x + vel_th * width_robot / 2.0;
    }
    vl = left_vel;
    vr = right_vel;
}
```

```
int main(int argc, char** argv){
    ros::init(argc, argv, "base_controller");
    ros::NodeHandle n;
    ros::Subscriber cmd_vel_sub = n.subscribe("cmd_vel", 10, cmd_
velCallback);
    ros::Rate loop_rate(10);

    while(ros::ok())
    {

        double dxy = 0.0;
        double dth = 0.0;
        ros::Time current_time = ros::Time::now();
        double dt;
        double velxy = dxy / dt;
        double velth = dth / dt;

      ros::spinOnce();
        dt =  (current_time - last_time).toSec();;
        last_time = current_time;

        // calculate odomety
        if(right_enc == 0.0){
          distance_left = 0.0;
          distance_right = 0.0;
        }else{
          distance_left = (left_enc - left_enc_old) / ticks_per_meter;
          distance_right = (right_enc - right_enc_old) / ticks_per_
meter;
        }

        left_enc_old = left_enc;
        right_enc_old = right_enc;

         dxy = (distance_left + distance_right) / 2.0;
         dth = (distance_right - distance_left) / width_robot;

        if(dxy != 0){
          x += dxy * cosf(dth);
          y += dxy * sinf(dth);
        }

        if(dth != 0){
          th += dth;
        }
        odom_quat = tf::createQuaternionMsgFromRollPitchYaw(0,0,th);
        loop_rate.sleep();
      }
}
```

> Notice that the equations are similar to diffdrive_plugin; this is because both robots are differential wheeled robots.

Do not forget to insert the following in your `CMakeLists.txt` file to create the executable of this file:

```
rosbuild_add_executable(base_controller src/base_controller.cpp)
```

This code is only a common example and must be extended with more code to make it work with a specific robot. It depends on the controller used, the encoders, and so on. We assume that you have the right background to add the necessary code in order to make the example work fine.

Creating a map with ROS

Getting a map can sometimes be a complicated task if you do not have the correct tools. ROS has a tool that will help you build a map using the odometry and a laser sensor. This tool is the **map_server** (`http://www.ros.org/wiki/slam_gmapping`). In this example, you will learn how to use the robot that we created in Gazebo, as we did in the previous chapters, to create a map, to save it, and load it again.

We are going to use a `.launch` file to make it easy. Create a new file in `chapter7_tutorials/launch` with the name `gazebo_mapping_robot.launch` and put in the following code:

```
<?xml version="1.0"?>
<launch>
  <param name="/use_sim_time" value="true" />

  <!-- start up wg world -->
  <include file="$(find gazebo_worlds)/launch/wg_collada_world.
launch"/>

  <arg name="model" />
  <param name="robot_description" command="$(find xacro)/xacro.py
$(arg model)" />

  <node name="spawn_robot" pkg="gazebo" type="spawn_model" args="-urdf
-param robot_description -z 0.1 -model robot_model" respawn="false"
output="screen" />

  <node name="rviz" pkg="rviz" type="rviz" args="$(find chapter7_
tutorials/)launch/mapping.vcg"/>
```

```
  <node name="slam_gmapping" pkg="gmapping" type="slam_gmapping">
  <remap from="scan" to="base_scan/scan"/>
  </node>
</launch>
```

With this `.launch` file, you can launch the Gazebo simulator with the 3D model, the `rviz` program with the correct configuration file, and `slam_mapping` to build a map in real time. Launch the file in a shell, and in the other shell, run the teleoperation node to move the robot:

```
$ roslaunch chapter7_tutorials gazebo_mapping_robot.launch
model:="`rospack find chapter7_tutorials`/urdf/robot1_base_04.xacro"
$ rosrun erratic_teleop erratic_keyboard_teleop
```

When you start to move the robot with the keyboard, you will see the free and the unknown space on the `rviz` screen as well as the map with the occupied space; this is known as an **Occupancy Grid Map** (**OGM**). The `slam_mapping` node updates the map state when the robot moves, or more specifically, when (after some motion) it has a good estimate of the robot's location and how the map is. It takes the laser scans and the odometry and builds the OGM for you.

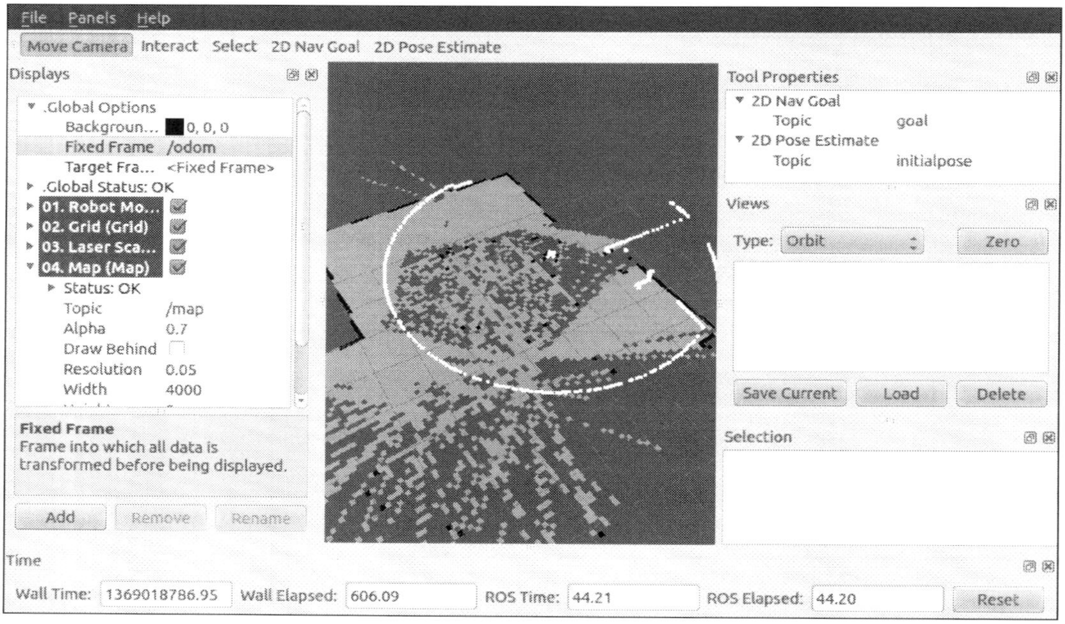

Saving the map using map_server

Once you have a complete map or something acceptable, you can save it to use it later in the navigation stack. To save it, use the following command:

```
$ rosrun map_server map_saver -f map
```

```
[ INFO] [1374457427.140784742]: Waiting for the map
[ INFO] [1374457427.875320754, 19.644000000]: Received a 4000 X 4000 map @ 0.050 m/pix
[ INFO] [1374457427.876429277, 19.644000000]: Writing map occupancy data to map.pgm
[ INFO] [1374457432.247321199, 20.040000000]: Writing map occupancy data to map.yaml
[ INFO] [1374457432.257071257, 20.040000000]: Done
```

This command will create two files, map.pgm and map.yaml. The first one is the map in the .pgm format (the portable gray map format). The other is the configuration file for the map. If you open it, you will see the following output:

```
image: map.pgm
resolution: 0.050000
origin: [-100.000000, -100.000000, 0.000000]
negate: 0
occupied_thresh: 0.65
free_thresh: 0.196
```

Now, open the .pgm image file with your favorite viewer, and you will see the map built before you:

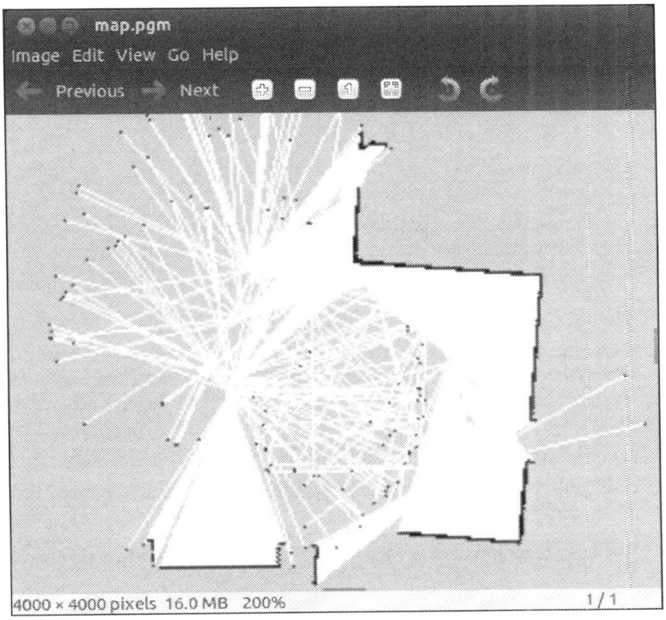

Loading the map using map_server

When you want to use the map built with your robot, it is necessary to load it with the `map_server` package. The following command will load the map:

```
$ rosrun map_server map_server map.yaml
```

But to make it easy, create another `.launch` file in `chapter7_tutorials/launch` with the name `gazebo_map_robot.launch` and put in the following code:

```
<?xml version="1.0"?>
<launch>
   <param name="/use_sim_time" value="true" />
    <!-- start up wg world -->
    <include file="$(find gazebo_worlds)/launch/wg_collada_world.
launch"/>
    <arg name="model" />
    <param name="robot_description" command="$(find xacro)/xacro.py
$(arg model)" />
    <node name="joint_state_publisher" pkg="joint_state_publisher"
type="joint_state_publisher" ></node>
    <!-- start robot state publisher -->
  <node pkg="robot_state_publisher" type="state_publisher"
name="robot_state_publisher" output="screen" >
  <param name="publish_frequency" type="double" value="50.0" />
  </node>
    <node name="spawn_robot" pkg="gazebo" type="spawn_model"
args="-urdf -param robot_description -z 0.1 -model robot_model"
respawn="false" output="screen" />
    <node name="map_server" pkg="map_server" type="map_server" args="
$(find chapter7_tutorials)/maps/map.yaml" />
    <node name="rviz" pkg="rviz" type="rviz" args="-d $(find chapter7_
tutorials)/launch/mapping.vcg" />
</launch>
```

And now, launch the file using the following command and remember to put the model of the robot that will be used:

```
$ roslaunch chapter7_tutorials gazebo_map_robot.launch model:="`rospack
find chapter7_tutorials`/urdf/robot1_base_04.xacro"
```

Then you will see `rviz` with the robot and the map. The navigation stack, in order to know the localization of the robot, will use the map published by the map server and the laser readings. This will help it perform a scan matching algorithm that helps to estimate the robot's location using a particle filter implemented in the `amcl` (**adaptive Monte Carlo localization**) node.

We will see more about maps as well as more useful tools in the next chapter.

Summary

In this chapter, you worked on the steps required to configure your robot in order to use it with the navigation stack. Now you know that the robot must have a planar laser, must be a differential wheeled robot, and it should satisfy some requirements for the base control and the geometry.

Keep in mind that we are working with Gazebo to demonstrate the examples and explain how the navigation stack works with different configurations. It is more complex to explain all of this directly on a real, robotic platform because we do not know whether you have one or have access to one. In any case, depending on the platform, the instructions may vary and the hardware may fail, so it is safer and useful to run these algorithms in simulations; later, we can test them on a real robot, as long as it satisfies the requirements described thus far.

In the next chapter, you will learn how to configure the navigation stack, create the `.launch` files, and navigate autonomously in Gazebo with the robot that you created in the previous chapters.

In brief, what you will learn after this chapter will be extremely useful because it shows you how to configure everything correctly so you know how to use the navigation stack with other robots, either simulated or real.

8
Navigation Stack – Beyond Setups

We are getting close to the end of the book, and this is when we will use all the knowledge acquired through it. We have created packages, nodes, 3D models of robots, and more. In *Chapter 7, Navigation Stack – Robot Setups* you configured your robot in order to be used with the navigation stack, and in this chapter, we will finish the configuration for the navigation stack so that you will learn how to use it with your robot.

All the work done in the previous chapters has been a preamble for this precise moment. This is when the fun begins and when the robots come alive.

In this chapter we are going to learn how to do the following:

- Apply the knowledge of *Chapter 7, Navigation Stack – Robot Setups* and the programs developed
- Understand the navigation stack and how it works
- Configure all the necessary files
- Create launch files to start the navigation stack

Let's go.

Creating a package

The correct way to do it is by adding the dependencies with the other packages created for your robot. For example, you could use the next command to create the package:

```
$ roscreate-pkg my_robot_name_2dnav move_base my_tf_configuration_dep my_
odom_configuration_dep my_sensor_configuration_dep
```

But in our case, as we have everything in the same package, it is only necessary to execute the following:

```
$ roscreate-pkg chapter8_tutorials roscpp
```

Remember that in the repository, you may find all the necessary files for the chapter.

Creating a robot configuration

To launch the entire robot, we are going to create a launch file with all the necessary files to activate all the systems.

> Keep in mind that this example is for a simulated robot in Gazebo. If you remember the previous chapters, Gazebo publishes the odometry and the tf for us.

Anyway, here you have a launch file for a real robot that you can use as a template.

The following script is present in `configuration_template.launch`:

```
<launch>
  <node pkg="sensor_node_pkg" type="sensor_node_type" name="sensor_
node_name" output="screen">
    <param name="sensor_param" value="param_value" />
  </node>

  <node pkg="odom_node_pkg" type="odom_node_type" name="odom_node"
output="screen">
    <param name="odom_param" value="param_value" />
  </node>

  <node pkg="transform_configuration_pkg" type="transform_
configuration_type" name="transform_configuration_name"
output="screen">
    <param name="transform_configuration_param" value="param_value" />
  </node>
</launch>
```

This launch file will launch three nodes that will start up the robot.

The first one is the node responsible for activating the sensors, for example, the **Laser Imaging, Detection, and Ranging (LIDAR)** system. The parameter `sensor_param` can be used to configure the sensor's port, for example, if the sensor uses a USB connection. If your sensor needs more parameters, you need to duplicate the line and add the necessary parameters. Some robots have more than one sensor to help in the navigation. In this case, you can add more nodes or create a launch file for the sensors and include it in this launch file. This could be a good option for easily managing all the nodes in the same file.

The second node is to start the odometry, the base control, and all the necessary files to move the base and calculate the robot's position. Remember that in *Chapter 7, Navigation Stack – Robot Setups* we did these nodes. Like in the other section, you can use the parameters to configure something in the `odometry` or replicate the line to add more nodes.

The third part is meant to launch the node responsible for publishing and calculating the geometry of the robot, and the transform between arms, sensors, and so on.

The previous file is for your real robot, but for our example, the next launch file is all we need.

Create a new file in `chapter8_tutorials/launch` with the name `chapter8_configuration_gazebo.launch` and add the following code:

```
<launch>

  <param name="/use_sim_time" value="true" />

  <!-- start up wg world -->
  <include file="$(find gazebo_worlds)/launch/wg_collada_world.
launch"/>

  <arg name="model" default="$(find chapter8_tutorials)/urdf/robot1_
base_04.xacro"/>
  <param name="robot_description" command="$(find xacro)/xacro.py
$(arg model)" />

  <node name="joint_state_publisher" pkg="joint_state_publisher"
type="joint_state_publisher" ></node>
  <!-- start robot state publisher -->
  <node pkg="robot_state_publisher" type="state_publisher"
name="robot_state_publisher" output="screen" />
```

```
  <node name="spawn_robot" pkg="gazebo" type="spawn_model" args="-urdf
-param robot_description -z 0.1 -model robot_model" respawn="false"
output="screen" />

  <node name="rviz" pkg="rviz" type="rviz" args="-d $(find chapter8_
tutorials)/launch/navigation.vcg" />

</launch>
```

This launch file is the same that we used in the previous chapters, so it does not need any additional explanation.

Now to launch this file, use the next command.

```
$ roslaunch chapter8_tutorials chapter8_configuration_gazebo.launch
```

You will see the next window:

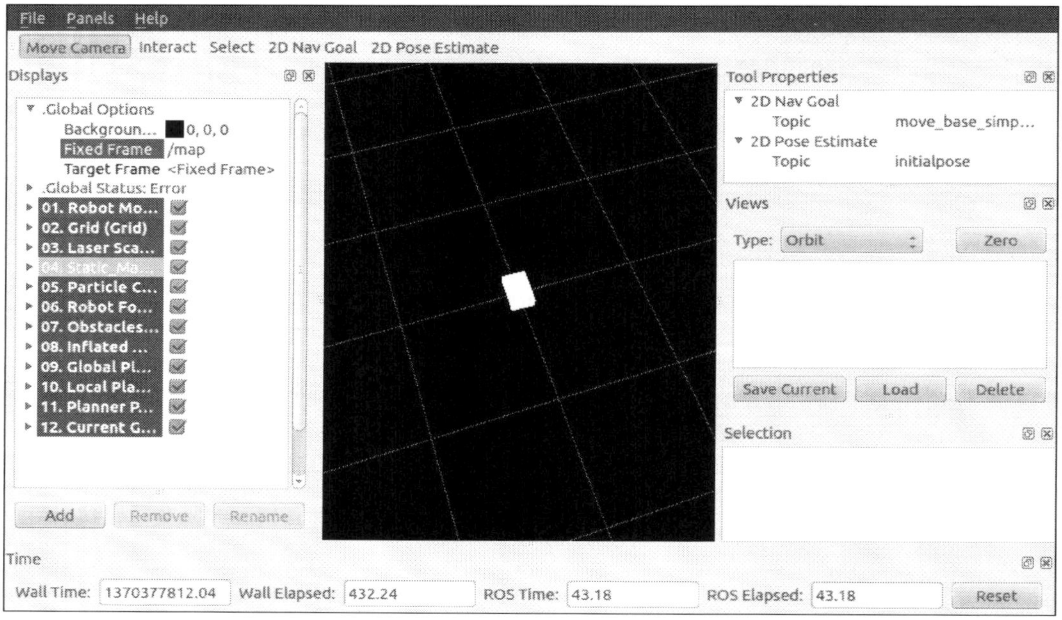

Notice that in the previous screenshot, there are some fields in red, blue, and yellow, without you having to configure anything before. This is because in the launch file, a configuration file for the `rviz` layout is loaded along with `rviz`, and this file was configured in the previous chapter of this book.

In the upcoming sections, you will learn how to configure `rviz` to use it with the navigation stack and view all the topics.

Configuring the costmaps (global_costmap) and (local_costmap)

Okay, now we are going to start configuring the navigation stack and all the necessary files to start it. To start with the configuration, first we will learn what costmaps are and what they are used for. Our robot will move through the map using two types of navigation—`global` and `local`.

- The `global` navigation is used to create paths for a goal in the map or a far-off distance
- The `local` navigation is used to create paths in the nearby distances and avoid obstacles, for example, a square window of 4 x 4 meters around the robot

These modules use costmaps to keep all the information of our map. The `global` costmap is used for the `global` navigation and the `local` costmap for `local` navigation.

The costmaps have parameters to configure the behaviors, and they have common parameters as well, which are configured in a shared file.

Configuration basically consists of three files where we can set up different parameters. The files are as follows:

- `costmap_common_params.yaml`
- `global_costmap_params.yaml`
- `local_costmap_params.yaml`

Just by reading the names of these configuration files, you can instantly guess what they are used for. Now that you have a basic idea about the usage of costmaps, we are going to create the configuration files and explain the parameters that are configured in them.

Configuring the common parameters

Let's start with the common parameters. Create a new file in `chapter8_tutorials/launch` with the name `costmap_common_params.yaml` and add the following code.

The following script is present in `costmap_common_params.yaml`:

```
obstacle_range: 2.5
raytrace_range: 3.0
```

```
footprint: [[-0.2,-0.2],[-0.2,0.2], [0.2, 0.2], [0.2,-0.2]]
#robot_radius: ir_of_robot
inflation_radius: 0.55

observation_sources: laser_scan_sensor

laser_scan_sensor: {sensor_frame: laser_base_link, data_type:
LaserScan, topic: /base_scan/scan, marking: true, clearing: true}
```

This file is used to configure common parameters. The parameters are used in `local_costmap` and `global_costmap`. Let's break the code and understand it.

The `obstacle_range` and `raytrace_range` attributes are used to indicate the maximum distance that the sensor will read and introduce new information in the costmaps. The first one is used for the obstacles. If the robot detects an obstacle closer than 2.5 meters in our case, it will put the obstacle in the costmap. The other one is used to clean/clear the costmap and update the free space in it when the robot moves. Note that we can only detect the "echo" of the laser or sonar with the obstacle, we cannot perceive the whole obstacle or object itself, but this simple approach will be enough to deal with these kinds of measurements, and we will be able to build a map and localize within it.

The `footprint` attribute is used to indicate to the navigation stack the geometry of the robot. It will be used to keep the right distance between the obstacles and the robot, or to know if the robot can go through a door. The `inflation_radius` attribute is the value given to keep a minimal distance between the geometry of the robot and the obstacles.

With the observation sources, you can set the sensors used by the navigation stack to get the data from the real world and calculate the path.

In our case, we are using a simulated LIDAR in Gazebo, but we can use a point cloud to do the same.

The next line will configure the sensor's frame and the uses of the data:

```
laser_scan_sensor: {sensor_frame: laser_base_link, data_type:
LaserScan, topic: /base_scan/scan, marking: true, clearing: true}
```

The laser configured in the previous line is used to add and clear obstacles in the costmap. For example, you could add a sensor with a wide range to find obstacles and another sensor to navigate and clear the obstacles. The topic's name is configured in this line; it is important to configure it well because the navigation stack could wait for another topic, all this while the robot is moving, and it can crash into a wall or an obstacle.

Configuring the global costmap

The next file for the configuration is the `global costmap` configuration file. Create a new file in `chapter8_tutorials/launch` with the name `global_costmap_params.yaml` and add the following code:

```
global_costmap:
  global_frame: /map
  robot_base_frame: /base_footprint
  update_frequency: 1.0
  static_map: true
```

The `global_frame` and the `robot_base_frame` attributes define the transformation between the map and the robot. This transformation is for the global costmap.

You can configure the frequency for the updates for the costmap. In this case, it is 1 Hz. The `static_map` attribute is used for the global costmap to know if a map or the map server is used to initialize the costmap. If you aren't using a static map, set this parameter to `false`.

Configuring the local costmap

The previous file is to configure the local costmap, create a new file in `chapter8_tutorials/launch` with the name `local_costmap_params.yaml` and add the following code:

```
local_costmap:
  global_frame: /map
  robot_base_frame: /base_footprint
  update_frequency: 1.0
  publish_frequency: 2.0
  static_map: true
  rolling_window: false
  width: 10.0
  height: 10.0
  resolution: 0.1
```

The `global_frame`, `robot_base_frame`, `update_frequency`, and `static_map` parameters are the same as described in the previous section, *Creating a robot configuration*. The `publish_frequency` parameter determines the frequency to publish information. The `rolling_window` parameter is used to keep the costmap centered on the robot when it is moving around the world.

You can configure the dimensions and the resolution of the costmap with the `width`, `height`, and `resolution` parameters. The values are given in meters.

Base local planner configuration

Once we have the costmaps configured, it is necessary to configure the base planner. The base planner is used to generate the velocity commands to move our robot. Create a new file in `chapter8_tutorials/launch` with the name `base_local_planner_params.yaml` and add the following code:

```
TrajectoryPlannerROS:
  max_vel_x: 1
  min_vel_x: 0.5
  max_rotational_vel: 1.0
  min_in_place_rotational_vel: 0.4

  acc_lim_th: 3.2
  acc_lim_x: 2.5
  acc_lim_y: 2.5

  holonomic_robot: false
```

The `config` file will set the maximum and minimum velocities for your robot. Also, the acceleration is set.

The `holonomic_robot` parameter is `true` if you are using a holonomic platform. In our case, our robot is based on a non-holonomic platform and the parameter is set to `false`. A **holonomic vehicle** is one that can move in all the configured space from any position. In other words, if the places where the robot can go are defined by any x and y values in the environment, this means that the robot can move there from any position. For example, if the robot can move forward, backward, and laterally, it is holonomic. A typical case of a non-holonomic vehicle is a car, as it cannot move laterally, and from a given position, there are many other positions (or poses) that are not reachable. Also, a differential platform is non-holonomic.

Downloading the example code

You can download the example code files for all Packt books you have purchased from your account at `http://www.packtpub.com`. If you purchased this book elsewhere, you can visit `http://www.packtpub.com/support` and register to have the files e-mailed directly to you. You can also download these code files from `https://github.com/AaronMR/Learning_ROS_for_Robotics_Programming`.

Creating a launch file for the navigation stack

Now, we have all the files created and the navigation stack is configured. To run everything, we are going to create a launch file. Create a new file in the chapter8_tutorials/launch folder and put the next code in a file with the name move_base.launch:

```
<launch>

    <!-- Run the map server -->
   <node name="map_server" pkg="map_server" type="map_server"
args="$(find chapter8_tutorials)/maps/map.yaml" output="screen"/>

  <include file="$(find amcl)/examples/amcl_diff.launch" >
  </include>

  <node pkg="move_base" type="move_base" respawn="false" name="move_
base" output="screen">
  <rosparam file="$(find chapter8_tutorials)/launch/costmap_common_
params.yaml" command="load" ns="global_costmap" />
  <rosparam file="$(find chapter8_tutorials)/launch/costmap_common_
params.yaml" command="load" ns="local_costmap" />
  <rosparam file="$(find chapter8_tutorials)/launch/local_costmap_
params.yaml" command="load" />
  <rosparam file="$(find chapter8_tutorials)/launch/global_costmap_
params.yaml" command="load" />
  <rosparam file="$(find chapter8_tutorials)/launch/base_local_
planner_params.yaml" command="load" />
  </node>
</launch>
```

Notice that in this file, we are launching all the files created before. We will launch a map server as well with a map that we created in *Chapter 7, Navigation Stack – Robot Setups* and the amcl node.

The amcl node that we are going to use is for differential robots because our robot is a differential robot. If you want to use amcl with holonomic robots, you will need to use the amcl_omni.launch file. If you want to use another map, go to *Chapter 7, Navigation Stack – Robot Setups* and create a new one.

Now launch the file and type the next command on a new shell. Recall that before you launch this file, you must launch the chapter8_configuration_gazebo.launch file.

```
$ roslaunch chapter8_tutorials move_base.launch
```

And you will see the following window:

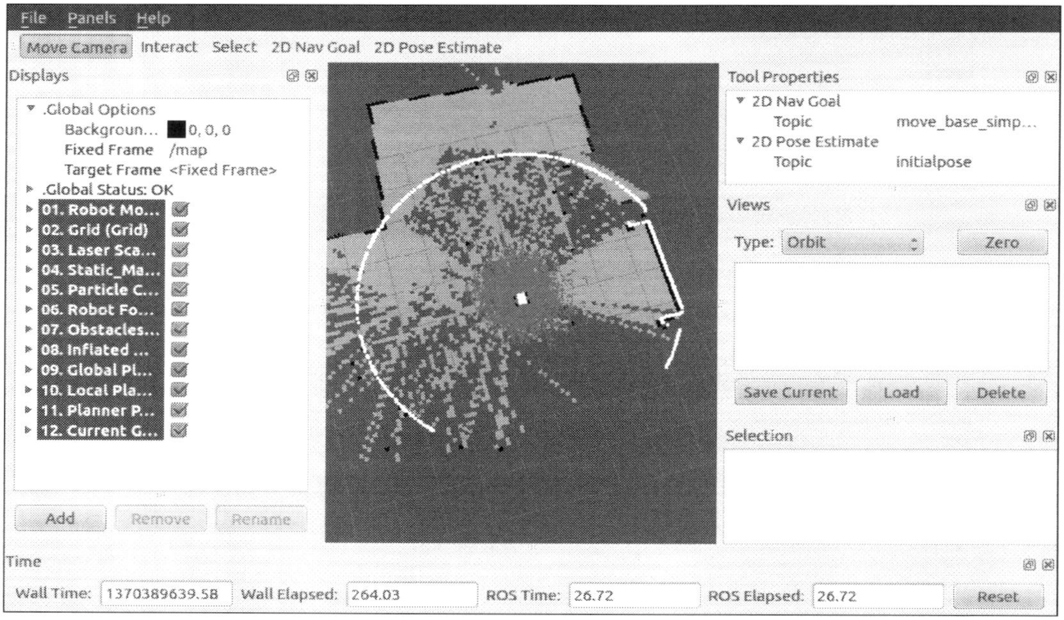

If you compare this image with the image that you saw when you launched the `chapter8_configuration_gazebo.launch` file, you will see that all the options are in blue; this is a good signal and it means that everything is OK.

As we said before, in the next section you will learn what options are necessary to visualize all the topics used in a navigation stack.

Setting up rviz for the navigation stack

It is good practice to visualize all the possible data to know what the navigation stack does. In this section, we will show you the visualization topic that you must add to `rviz` to see the correct data sent by the navigation stack. Discussions of each visualization topic that the navigation stack publishes is explained next.

2D pose estimate

The 2D pose estimate (P shortcut) allows the user to initialize the localization system used by the navigation stack by setting the pose of the robot in the world. The navigation stack waits for the new pose of a new topic with the name `initialpose`. This topic is sent using the rviz windows where we previously changed the name of the topic.

You can see in the following screenshot how you can use `initialpose`. Click on the **2D Pose Estimate** button and click on the map to indicate the initial position of your robot. If you don't do this at the beginning, the robot will start the auto-localization process and try to set an initial pose.

- `Topic: initialpose`
- `Type: geometry_msgs/PoseWithCovarianceStamped`

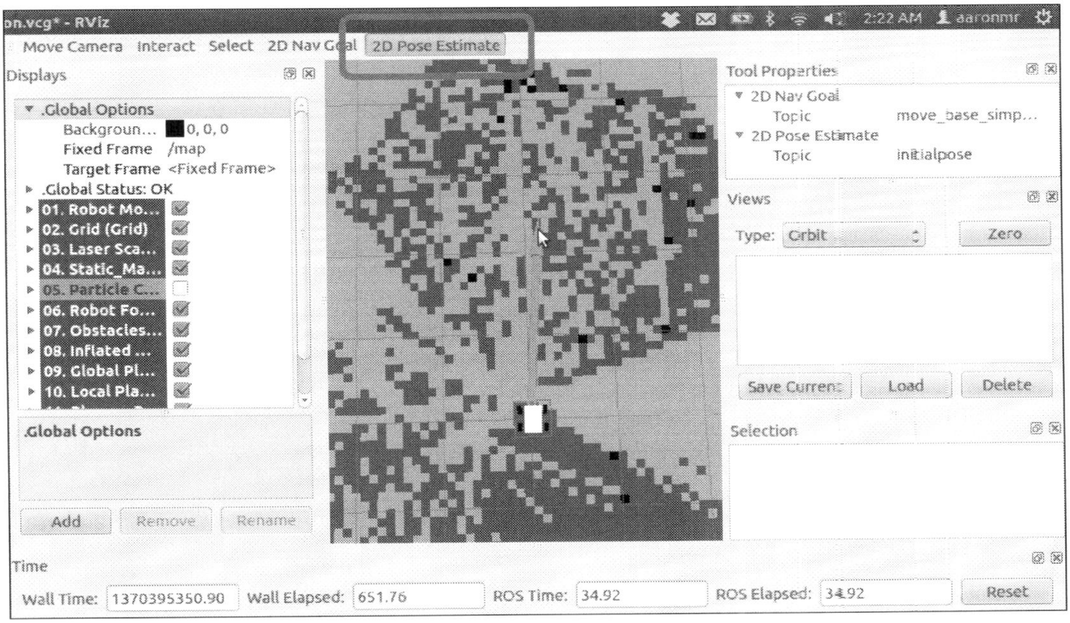

2D nav goal

The 2D nav goal (G shortcut) allows the user to send a goal to the navigation by setting a desired pose for the robot to achieve. The navigation stack waits for a new goal with the topic's name `/move_base_simple/goal`; for this reason, you must change the topic's name in the `rviz` windows in **Tool Properties** in the **2D Nav Goal** menu. The new name that you must put in this textbox is `/move_base_simple/goal`. In the next window, you can see how to use it. Click on the **2D Nav Goal** button and select the map and the goal for your robot. You can select the x and y position and the end orientation for the robot.

- `Topic: move_base_simple/goal`
- `Type: geometry_msgs/PoseStamped`

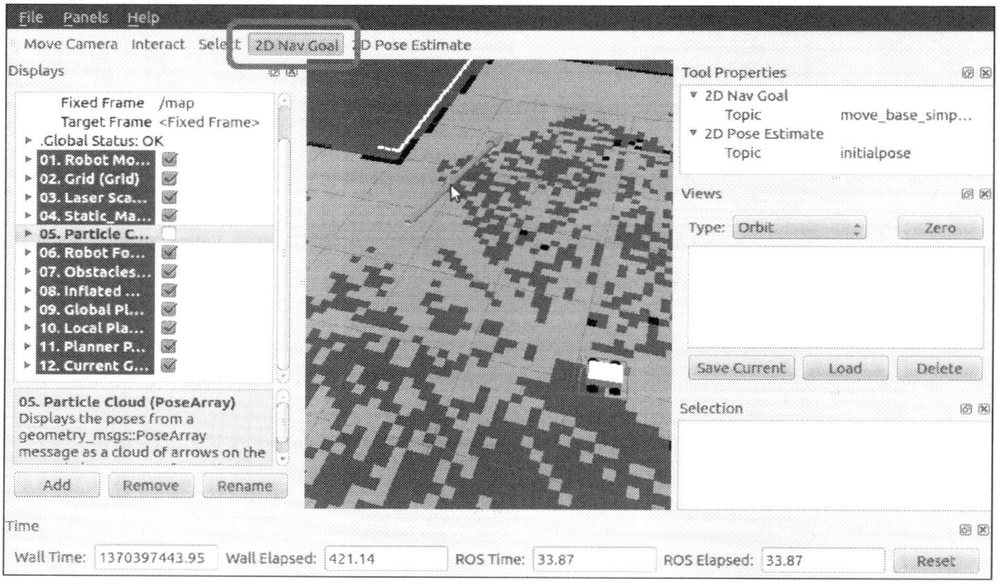

Static map

This displays the static map that is being served by the map_server, if one exists. When you add this visualization, you will see the map we captured in *Chapter 7, Navigation Stack – Robot Setups* in the *Creating a map with ROS* section.

In the next window, you can see the display type that you need to select and the name that you must put in the display name.

- `Topic: map`
- `Type: nav_msgs/GetMap`

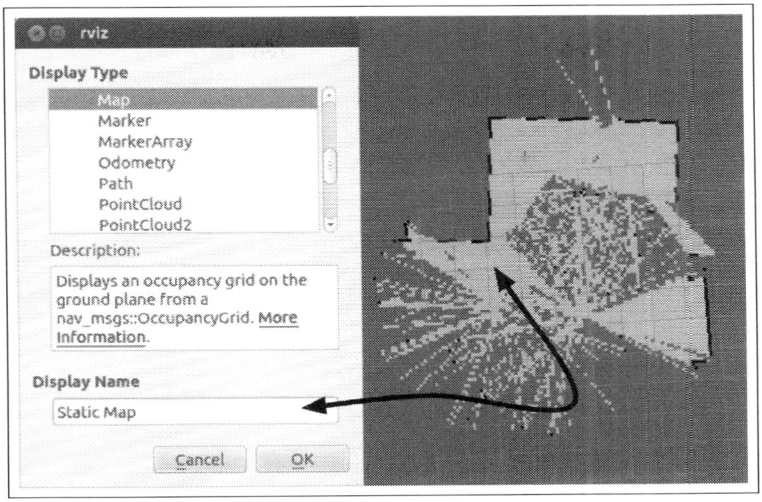

Particle cloud

It displays the particle cloud used by the robot's localization system. The spread of the cloud represents the localization system's uncertainty about the robot's pose. A cloud that spreads out a lot reflects high uncertainty, while a condensed cloud represents low uncertainty. In our case, you will obtain the next cloud for the robot:

- `Topic: particlecloud`
- `Type: geometry_msgs/PoseArray`

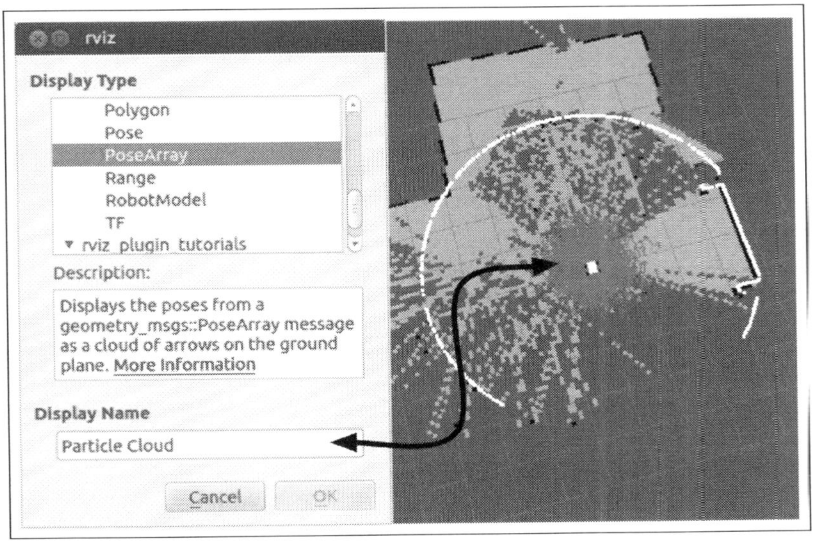

Robot footprint

It shows the footprint of the robot; in our case, the robot has a footprint, which has a width of 0.4 meters and a height of 0.4 meters. Remember that this parameter is configured in the `costmap_common_params` file. This dimension is important because the navigation stack will move the robot in a safe mode using the values configured before.

- `Topic: local_costmap/robot_footprint`
- `Type: geometry_msgs/Polygon`

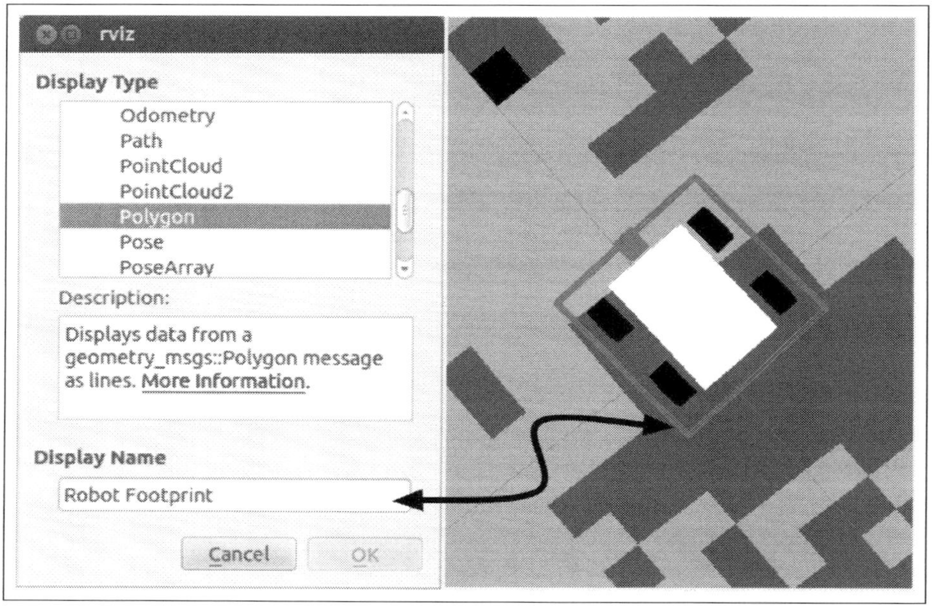

Obstacles

It shows the obstacles that the navigation stack sees in its costmap. In the following screenshot, you can see it in red over the black line in the map. The red line is the detected obstacle. You could also see red lines in the middle of the map if a temporal obstacle is in front of the robot. For the robot to avoid collision, the robot's footprint should never intersect with a cell that contains an obstacle.

- `Topic: local_costmap/obstacles`
- `Type: nav_msgs/GridCells`

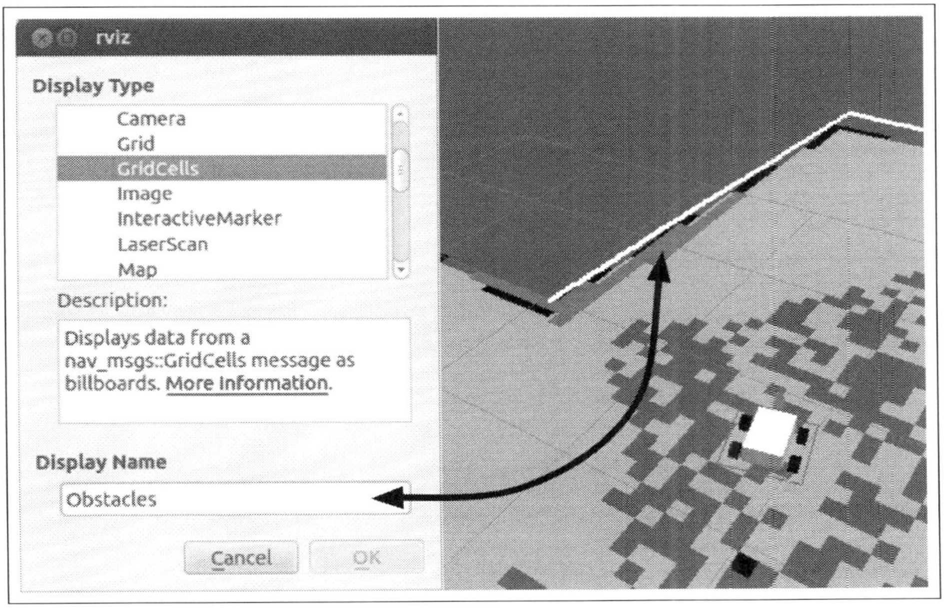

Inflated obstacles

It shows obstacles in the navigation stack's costmap inflated by the inscribed radius of the robot. Like the red line from the obstacle, you could see in the middle of the map blue lines or shapes that indicate that a temporal obstacle is in front of the robot. For the robot to avoid collision, the center point of the robot should never overlap with a cell that contains an inflated obstacle.

- `Topic: local_costmap/inflated_obstacles`
- `Type: nav_msgs/GridCells`

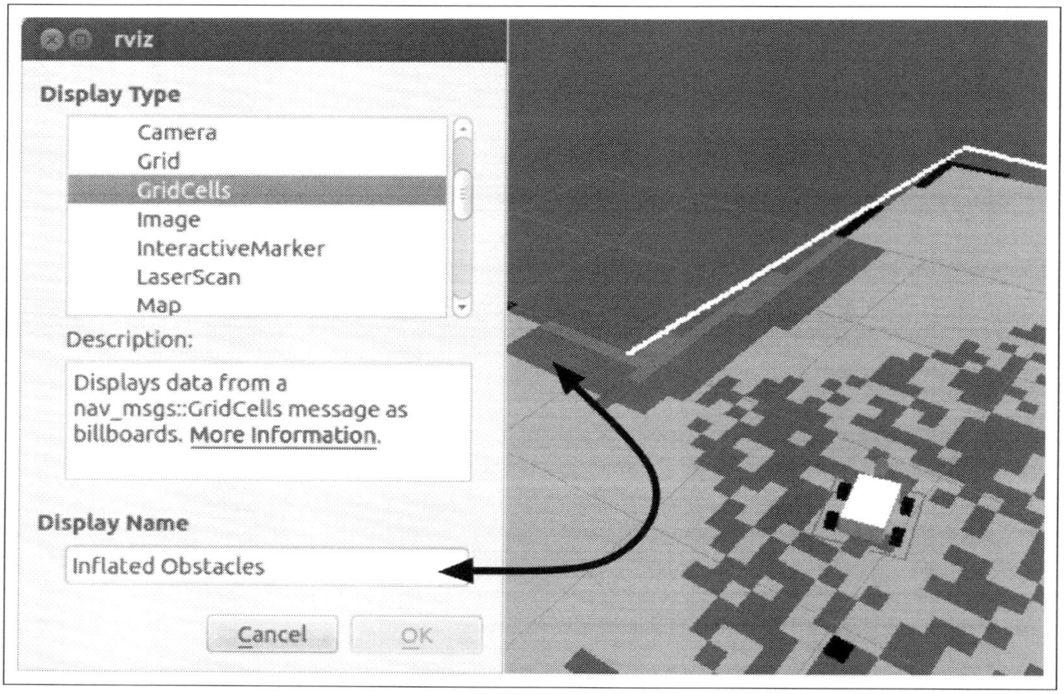

Global plan

It shows the portion of the global plan that the local planner is currently pursuing. You can see it in green in the next image. Perhaps the robot will find obstacles during the movement and the navigation stack will recalculate a new path to avoid collisions and try to follow the global plan.

- `Topic: TrajectoryPlannerROS/global_plan`
- `Type: nav_msgs/Path`

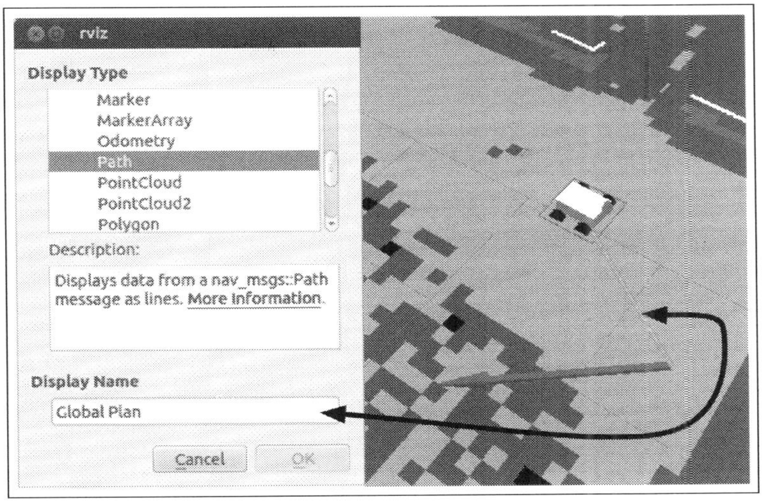

Local plan

It shows the trajectory associated with the velocity commands currently being commanded to the base by the local planner. You can see the trajectory in blue in front of the robot in the next image. You can use this display to know whether the robot is moving and the approximate velocity depending on the length of the blue line.

- Topic: `TrajectoryPlannerROS/local_plan`
- Type: `nav_msgs/Path`

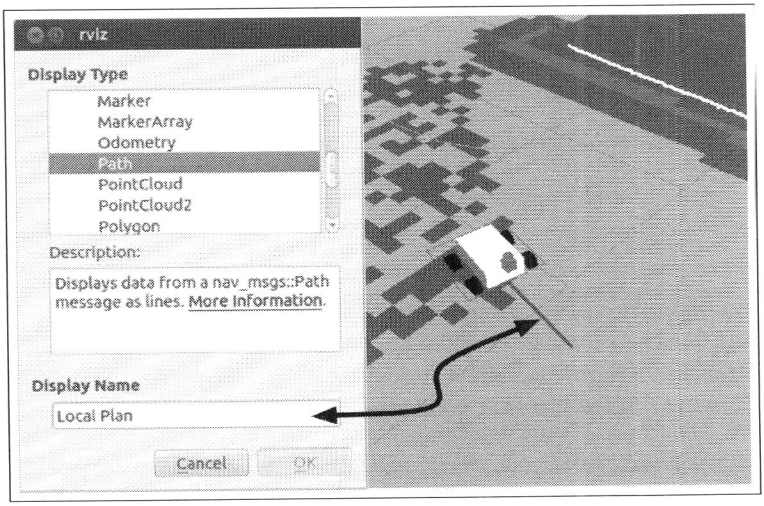

Planner plan

It displays the full plan for the robot computed by the global planner. You will see that it is similar to the global plan.

- `Topic: NavfnROS/plan`
- `Type: nav_msgs/Path`

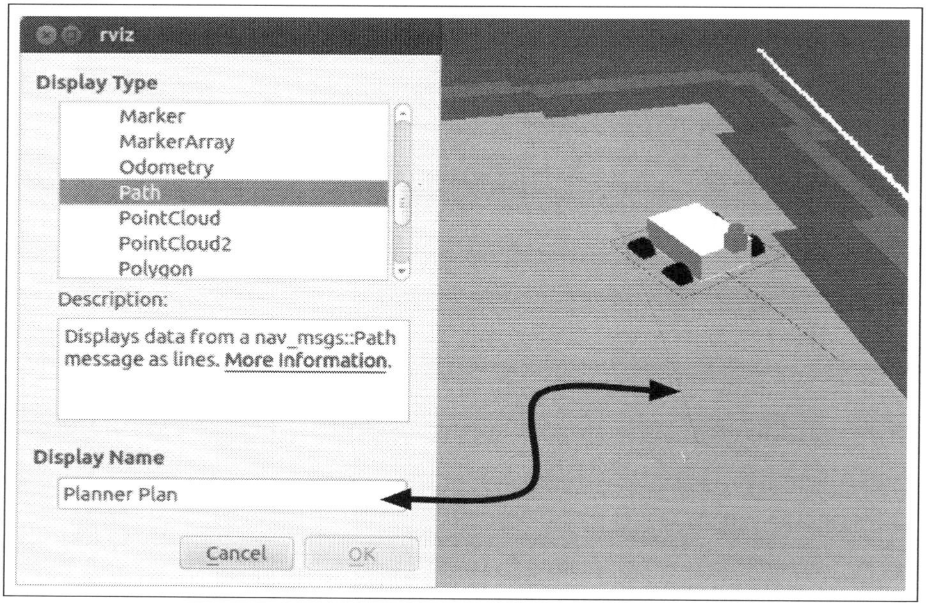

Current goal

It shows the goal pose that the navigation stack is attempting to achieve. You can see it as a red arrow, and it is displayed after you put a new 2D nav goal. It can be used to know the final position of the robot.

- `Topic: current_goal`
- `Type: geometry_msgs/PoseStamped`

These visualizations are all you need to see the navigation stack in rviz. With this, you can notice if the robot is doing something strange. Now we are going to see a general image of the system. Run `rxgraph` to see whether all the nodes are running and to see the relations between them.

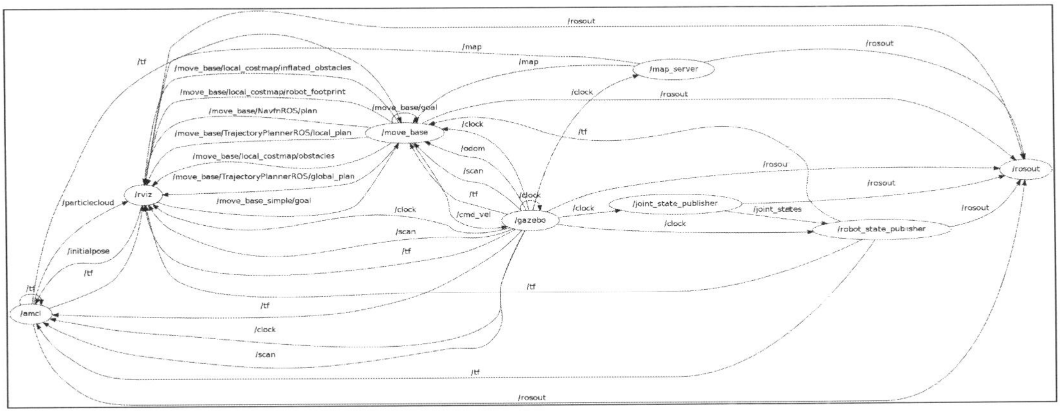

Adaptive Monte Carlo Localization (AMCL)

In this chapter, we are using the `amcl` algorithm for the localization. `amcl` is a probabilistic localization system for a robot moving in 2D. This system implements the adaptive Monte Carlo localization approach, which uses a particle filter to track the pose of a robot against a known map.

`amcl` has many configuration options that will affect the performance of localization. For more information on `amcl`, please see the AMCL documentation. In the following links, you can find more information about it:

- `http://www.ros.org/wiki/amcl`
- `http://www.probabilistic-robotics.org/`

The `amcl` node works mainly with laser scans and laser maps, but it could be extended to work with other sensor data, such as a sonar or stereo vision. So for this chapter, it takes a laser-based map and laser scans, and transforms messages and generates a probabilistic pose. On startup, `amcl` initializes its particle filter according to the parameters provided in the setup. If you don't set the initial position, `amcl` will start in the origin of the coordinates. Anyway, you can set the initial position in `rviz` using the **2D Pose Estimate** button.

When we include the `amcl_diff.launch` file, we are starting the node with a series of configured parameters. This configuration is the default configuration and the minimum setting to make it work.

Next, we are going to see the content of the `amcl_diff.launch` launch file to explain some parameters:

```
<launch>
<node pkg="amcl" type="amcl" name="amcl" output="screen">
  <!-- Publish scans from best pose at a max of 10 Hz -->
  <param name="odom_model_type" value="diff"/>
  <param name="odom_alpha5" value="0.1"/>
  <param name="transform_tolerance" value="0.2" />
  <param name="gui_publish_rate" value="10.0"/>
  <param name="laser_max_beams" value="30"/>
  <param name="min_particles" value="500"/>
  <param name="max_particles" value="5000"/>
  <param name="kld_err" value="0.05"/>
  <param name="kld_z" value="0.99"/>
  <param name="odom_alpha1" value="0.2"/>
  <param name="odom_alpha2" value="0.2"/>
```

```
    <!-- translation std dev, m -->
    <param name="odom_alpha3" value="0.8"/>
    <param name="odom_alpha4" value="0.2"/>
    <param name="laser_z_hit" value="0.5"/>
    <param name="laser_z_short" value="0.05"/>
    <param name="laser_z_max" value="0.05"/>
    <param name="laser_z_rand" value="0.5"/>
    <param name="laser_sigma_hit" value="0.2"/>
    <param name="laser_lambda_short" value="0.1"/>
    <param name="laser_lambda_short" value="0.1"/>
    <param name="laser_model_type" value="likelihood_field"/>
    <!-- <param name="laser_model_type" value="beam"/> -->
    <param name="laser_likelihood_max_dist" value="2.0"/>
    <param name="update_min_d" value="0.2"/>
    <param name="update_min_a" value="0.5"/>
    <param name="odom_frame_id" value="odom"/>
    <param name="resample_interval" value="1"/>
    <param name="transform_tolerance" value="0.1"/>
    <param name="recovery_alpha_slow" value="0.0"/>
    <param name="recovery_alpha_fast" value="0.0"/>
  </node>
  </launch>
```

The `min_particles` and `max_particles` parameters set the minimum and maximum number of particles that are allowed for the algorithm. With more particles, you get more accuracy, but this increases the use of the CPU.

The `laser_model_type` parameter is used to configure the laser type. In our case, we are using a `likelihood_field` parameter but the algorithm can also use beam lasers.

The `laser_likelihood_max_dist` parameter is used to set the maximum distance to do obstacle inflation on the map, which is used in the `likelihood_field` model.

The `initial_pose_x`, `initial_pose_y`, and `initial_pose_a` parameters are not in the launch file, but they are interesting because they set the initial position of the robot when the `amcl` starts, for example, if your robot always starts in the dock station and you want to set the position in the launch file.

Perhaps you should change some parameters to tune your robot and make it work fine. In the `ros.org` page, you have a lot of information about the configuration and the parameters that you could change.

Avoiding obstacles

A great functionality of `navigation stack` is the recalculation of the path if it finds obstacles during the movement. You can easily see this feature by adding an object in front of the robot in Gazebo. For example, in our simulation we added a big box in the middle of the path. The navigation stack detects the new obstacle and automatically creates an alternative path.

In the next image, you can see the object that we added. Gazebo has some predefined 3D objects that you can use in the simulations with mobile robots, arms, humanoids, and so on.

To see the list, go to the **Insert model** section. Select one of the objects and then click where you want to put it.

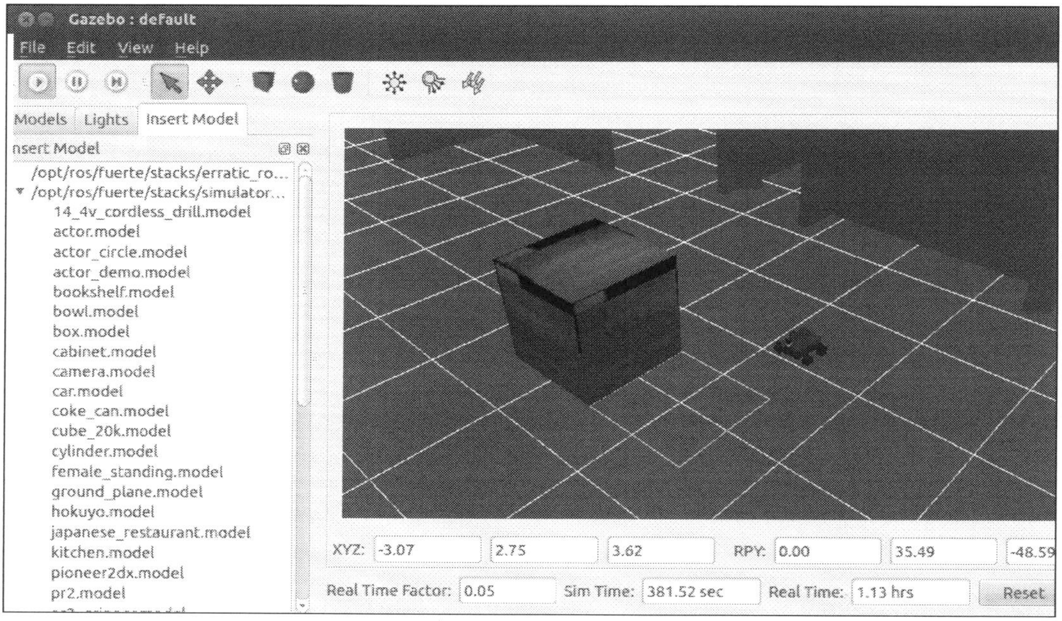

If you go to the `rviz` windows now, you will see a new global plan to avoid the obstacle. This feature is very interesting when you use the robot in real environments with people walking around the robot. If the robot detects a possible collision, it will change the direction and it will try to arrive at the goal. You can see this feature in the next image. Recall that the detection of such obstacles is reduced to the area covered by the local planner costmap (for example, 4x4 meters around the robot).

Sending goals

We are sure that you have been playing with the robot by moving it around the map a lot. This is funny but a little tedious and it is not very functional.

Perhaps you were thinking that it would be a great idea to program a list of movements and send the robot to different positions with only a button, even when we are not in front of a computer with `rviz`.

Okay, now you are going to learn how to do it using `actionlib`.

The `actionlib` package provides a standardized interface for interfacing with tasks. For example, you can use it to send goals for the robot to detect something in a place, make scans with the laser, and so on. In this section, we will send a goal to the robot and we will wait for this task to end.

It could look similar to services, but if you are doing a task that has a long duration, you might want the ability to cancel the request during the execution or get periodic feedback about how the request is progressing, and you cannot do it with services. Furthermore, `actionlib` creates messages (not services), and it also creates topics, so we can still send the goals through a topic without taking care of the feedback and result, if we do not want.

The next code is a simple example to send a goal to move the robot. Create a new file in the `chapter8_tutorials/src` folder and add the following code in a file with the name `sendGoals.cpp`:

```
#include <ros/ros.h>
#include <move_base_msgs/MoveBaseAction.h>
#include <actionlib/client/simple_action_client.h>
#include <tf/transform_broadcaster.h>
#include <sstream>

typedef actionlib::SimpleActionClient<move_base_msgs::MoveBaseAction>
MoveBaseClient;

int main(int argc, char** argv){
  ros::init(argc, argv, "navigation_goals");

  MoveBaseClient ac("move_base", true);

  while(!ac.waitForServer(ros::Duration(5.0))){
    ROS_INFO("Waiting for the move_base action server");
  }

  move_base_msgs::MoveBaseGoal goal;

  goal.target_pose.header.frame_id = "map";
  goal.target_pose.header.stamp = ros::Time::now();

  goal.target_pose.pose.position.x = 1.0;
  goal.target_pose.pose.position.y = 1.0;
  goal.target_pose.pose.orientation.w = 1.0;

  ROS_INFO("Sending goal");
  ac.sendGoal(goal);
```

```
  ac.waitForResult();

  if(ac.getState() == actionlib::SimpleClientGoalState::SUCCEEDED)
    ROS_INFO("You have arrived to the goal position");
  else{
    ROS_INFO("The base failed for some reason");
 }
  return 0;
}
```

Before we compile, add the following dependencies into the manifest file:

```
<depend package="move_base_msgs" />
<depend package="actionlib" />
<depend package="geometry_msgs" />
<depend package="tf" />
```

Add the next file in the `CMakeList.txt` file to generate the executable for our program:

```
rosbuild_add_executable(sendGoals src/sendGoals.cpp)
```

Now, compile the package with the following command:

```
$ rosmake chapter8_tutorials
```

And launch everything to test the new program. Use the next command to launch all the nodes and the configurations:

```
$ roslaunch chapter8_tutorials chapter8_configuration_gazebo.launch
$ roslaunch chapter8_tutorials move_base.launch
```

Once you have configured the 2D pose estimate, run the `sendGoal` node with the next command in a new shell:

```
$ rosrun chapter8_tutorials sendGoal
```

If you go to the `rviz` screen, you will see a new **Global Plan** (green line) over the map. This means that the navigation stack has accepted the new goal and it will start to execute it.

When the robot arrives at the goal, you will see the next message on the shell where you ran the node:

```
[ INFO ] [...,...]: You have arrived to the goal position
```

You can do a list of goals or waypoints and make a route for the robot. This way you can program missions, guardian robots, or bring things from other rooms with your robot.

Summary

At the end of this chapter, you should have a robot—simulated or real—moving autonomously through the map (which models the environment) using the navigation stack. You can program the control and the localization of the robot by following the ROS philosophy of code reusability, so that you can have the robot completely configured without much effort. The most difficult part of this chapter is to understand all the parameters and learn how to use each one of them appropriately. The correct use of them will determine whether your robot works fine or not; for this reason, you must practice changing the parameters and look for the reaction of the robot.

In the next chapter, you will see some robots that you can download and simulate on ROS. These robots have 3D models and they can be used in Gazebo for simulation. Some of them have the navigation stack implemented, and you will learn how to use it and play with them. It should not be a problem for you because you have all the knowledge to program something similar and launch the system and localize errors and fix them.

9

Combining Everything – Learn by Doing

We will finish this book by showing how to work with real robots, which are regularly used by research groups and robotic companies worldwide. Several of these groups and companies have already ported the software of their robots to the ROS framework, and they have made the ROS packages for their robots publicly available. As we will see in the following sections, these packages usually comprise several stacks, covering the navigation, the robot-specific drivers, teleoperation, and the robot URDF model for the simulation in Gazebo.

Here we will show you how to install the packages for the most known robots, and work with them in simulation. Note that most of these robots are very expensive, but you can develop for them and contribute your algorithms to the manufacturers. Anyway, in some cases the robots are affordable by the amateur user, like in the case of the turtlebot based on the roomba. Nevertheless, you will be able to run and work with all the robots listed here with a single PC and your ROS installation with the navigation stack and Gazebo.

For each robot covered in this chapter, we will include a brief description of the robot features, the company that develops it, and the links to the sources, which are the ROS packages. Then, we will show the installation steps, probably with some advice for a given ROS distribution. Once the system is installed, we will run a few examples of the robot running and doing some ground-breaking stuff. In all cases, the "hello world" example will consist of loading the URDF model of the robot on Gazebo. Then, we will show how to move it, navigate it in the simulated environment, and access its simulated sensors and actuators, specially manipulators and arms.

The outline of this chapter will cover the following robots and demos:

- The REEM from PAL Robotics, in particular the REEM-H, is a wheeled differential platform humanoid robot. We will also introduce other robots from this company, such as the biped REEM-C. We will show how to install the public code of these robots and navigate the REEM-H in simulation.
- The PR2 from Willow Garage, the original maintainers and developers of ROS. We will see how to install the robot along with the official repositories and how to navigate it in Gazebo, selecting different parts of the vehicle. We will also show how to build a map in order to do mapping as we showed in the previous two chapters. Finally, some demos are enumerated.
- The Robonaut 2, or simply R2, is a dexterous humanoid of NASA. We will show how to install, load the robot in simulation, move the arms, and also play with several options to see the legs, the ISS environment, and much more.
- The Husky rover from Clearpath Robotics is a 4-wheel outdoors ground vehicle. We will show how to install and run it in simulation.
- We will finish with the TurtleBot, a low-cost mobile robot of Yujin Robot and iRobot, which is one of the most affordable vehicles to start with in robotics. We will see how to run it inside Gazebo, after installing from the official repository.

You will see that for some robots we will advise to use ROS Groovy instead of ROS Fuerte, which has been the distribution used through the entire book. We have already used Groovy for some particular parts, because they were quite recent and better supported in Groovy. Here, the same thing happens. For many robots, the best support is found in Groovy, although in Fuerte you might also be able to run them. Anyway, you already know how to install ROS, and installing Groovy is the same as Fuerte, as we explained in the first chapter; you only have to select `ros-groovy-*` in the Debian packages of Ubuntu. Moreover, you also know how to call the `setup.bash` file of the distro you want to use from the `.bashrc` file, as you can actually have both distribution packages installed in your machine.

REEM – the humanoid of PAL Robotics

REEM is a 1.65m humanoid service robot created by PAL Robotics. This robot is one of several humanoid robots developed by the company. In brief, it designs two kinds of humanoids: bipeds and differential drive robots. REEM is the last version of their differential driver robots and it is one of the state-of-the-art robots of its kind. It has an autonomous navigation system, a user-friendly touchscreen on the chest for **Human Robot Interaction** (**HRI**), and it incorporates a voice and face recognition system.

REEM is meant to help and entertain people in most public environments such as hotels, museums, shopping malls, airports, and hospitals. It can also transport small packages, and operate as a dynamic information point offering a great variety of multimedia applications such as display an interactive map of the surrounding area, call up a variety of information (weather, nearby restaurants, airlines' travel schedule, and so on), and offer tele-assistance via video-conferencing, according to their website `http://pal-robotics.com/robots`.

PAL Robotics (`http://pal-robotics.com`) is a company under the umbrella of the Abu Dhabi based Royal Group. It is an R&D company with the ultimate goal of building the ideal robot for everybody's use. Their first robot was the biped robot REEM-A, that was finished in 2005. It participated in the 2006 RoboCup competition and won the walking competition. REEM-B is the second biped developed by the company and recently they have developed REEM-C. All these three robots are shown from the left- to the right-hand side in the following picture. They have also designed two differential drive humanoids—with a mobile wheel base— REEM-H1 being the first of them.

REEM-H1, shown in the following picture (on the right-hand side), is the predecessor of the last designed robot called REEM, as shown in the previous image.

You can see the robots and products of PAL Robotics on their homepage, as well as the news and events of the company such as public demos or conferences, on their blog—`http://www.pal-robotics.com/blog/`. Similarly, they offer the software as a set of open source ROS packages: `http://pal-robotics.com/software`. From this link of the software section, we jump to the GitHub repository of PAL Robotics robots; GitHub is a web host that hosts several git repositories, which offers easy access to them. In the next section we explain how to install it and what they provide in each stack hosted in the repository.

Installing REEM from the official repository

If you visit the link `https://github.com/pal-robotics`, you will see a number of stacks that comprise the whole system to integrate the REEM robot in ROS. The easiest way to install all the stacks is by using the `rosinstall` files that are inside the `pal-ros-pkg` repository. Before that, we must install some dependencies. In brief, we need the standalone version of the Gazebo simulator, as the one that comes with ROS is older than the one used by REEM. To install the standalone version of `gazebo`, we must add the Ubuntu repositories that host it. As it is maintained by the OSRF, we set up the `packages.osrfoundation.org` sources as follows (select your Ubuntu distro accordingly and note that we use `quantal` in this example):

```
sudo sh -c 'echo "deb http://packages.osrfoundation.org/gazebo/ubuntu
quantal main" > /etc/apt/sources.list.d/gazebo-latest.list'
```

```
wget http://packages.osrfoundation.org/gazebo.key -O - | sudo apt-key add
-
```

Update the system package list as follows:

```
sudo apt-get update
```

And install Gazebo as shown:

```
sudo apt-get install gazebo
```

Finally, source the `setup.sh` of `gazebo` in your `.bahrc` file:

```
echo "source /usr/share/gazebo/setup.sh" >> ~/.bashrc
source ~/.bashrc
```

For your convenience, we have all these steps in the `scripts` folder of the code that comes with this chapter. Just run the following to install Gazebo:

```
./gazebo_install.sh
```

The second dependency is actually part of the `drcsim` simulator, but we do not need it completely. We only need the OSRF common libraries. In order to install them, move to your `ROS_PACKAGE_PATH` package and download the source from their repository with another control version system, such as `cvs`, `subversion`, or `git` (note that you need mercurial in order to use `hg`).

```
hg clone https://bitbucket.org/osrf/osrf-common
```

Now build and install it system-wide as follows:

```
cd osrf-common
mkdir build
cd build
cmake ..
make
sudo make install
```

Again, for your convenience, all these steps are accomplished by the next script:

```
./osrf_common_install.sh
```

The rest of the dependencies can be installed easily with `apt-get`:

```
sudo apt-get install ros-groovy-pr2-simulator
sudo apt-get install libglew-dev libdevil-dev
```

Note that we have used ROS Groovy in the previous commands. ROS Fuerte should work as well, but the developers of REEM are currently porting the source code to Groovy in this repository, so it is safer to use Groovy, as otherwise some parts might not compile or may crash.

Basically, we need the simulator libraries that come for the PR2, but we do not need the PR2 at all, only those libraries.

At this point, we must add some environment variables to our `.bashrc` file. Just copy this at the end of the file:

```
# Gazebo - atlas_msgs; not required if drcsim is installed
export GAZEBO_PLUGIN_PATH="`rospack find atlas_msgs`/lib:$GAZEBO_PLUGIN_
PATH"
export LD_LIBRARY_PATH="`rospack find atlas_msgs`/lib:$LD_LIBRARY_PATH"

# Gazebo - PAL
export GAZEBO_PLUGIN_PATH="`rospack find pal_gazebo_plugins`/lib:$GAZEBO_
PLUGIN_PATH"
export LD_LIBRARY_PATH="`rospack find pal_gazebo_plugins`/lib:$LD_
LIBRARY_PATH"

# Gazebo - REEM-H3
export GAZEBO_MODEL_PATH="`rospack find reem_gazebo`/models:$GAZEBO_
MODEL_PATH"
export GAZEBO_RESOURCE_PATH="`rospack find reem_gazebo`/worlds:$GAZEBO_
RESOURCE_PATH"
```

For your convenience, these environment variables are in the file `environment/pal_ros_pkg_environment.sh`, so just do as follows:

```
cat scripts/environment/pal_ros_pkg_environment.sh >> ~/.bashrc
```

Now, we have all the dependencies installed and the environment correctly set up. Move to your `ROS_PACKAGE_PATH` and create a new folder that will contain all the stacks for the REEM robot. We will use the `ros-pal-pkg` folder. Go inside this directory and just run the following:

```
rosinstall . https://raw.github.com/pal-robotics/pal-ros-pkg/master/pal-
ros-pkg-gazebo-standalone.rosinstall
```

This will download all the repositories containing the stacks for REEM; this folder (ros-pal-pkg) can be omitted but this way it will be easier to know which are the repositories (and the stacks and packages) of REEM, because there are quite a lot. Note that we are using the Gazebo-standalone rosintall file, since we have already installed gazebo as a standalone package on the system; there is a version that tries to use the Gazebo binaries that comes with the ROS distro, but it is unstable.

Once everything has been downloaded, we must update the environment before we install anything, so just do as follows:

```
rosstack profile && rospack profile
source ~/.bashrc
```

Now we can install every stack, one after the other. You can do it with this bash snippet:

```
stacks=(atlas_msgs pal_gazebo_plugins pal_image_processing perception_
blort reem_arm_navigation reem_common reem_controllers reem_kinematics
reem_msgs reem_robot reem_simulation)


for stack in ${stacks[@]}
do
  rosmake -i $stack
done
```

Now you have the REEM software installed on the system. All the previous commands are compiled in a script. Although it is not intended for system-wide installation, you could just run the following:

```
./pal_ros_pkg_install.sh
```

From all the stacks installed, the most important ones are reem_simulation, reem_teleop and reem_arm_navigation, which allow simulating the robot in Gazebo and teleoperating the base and the arms.

Running REEM using the Gazebo simulator

The REEM sources include a launch file to load it in Gazebo. You only have to run the following code and you will see REEM in an empty environment in Gazebo, as shown in the following figure that also includes the node graph using `rxgraph`:

```
roslaunch reem_gazebo reem_empty_world.launch
```

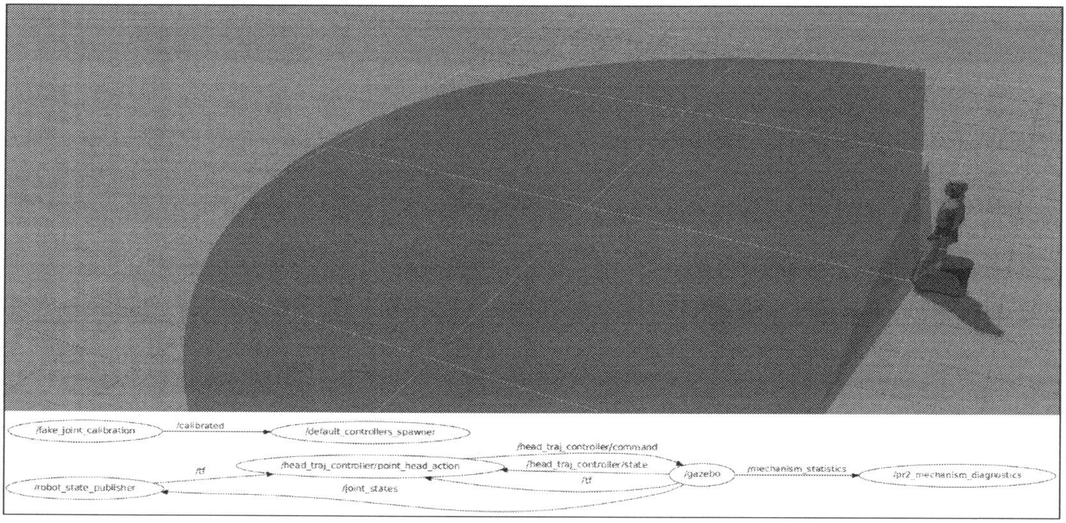

In the screenshot you can observe REEM has two frontal range lasers. One pointing forward with a large range and another slightly tilted down. We could load any environment we want in Gazebo and use the navigation stack to make the robot move through the simulated environment, as discussed in previous chapters.

If you experience any problem when launching REEM in Gazebo, it may be related to the version of Gazebo being used. In that case, try some of the precompiled binaries of Gazebo available at `http://gazebosim.org/assets/distributions/`; for instance, Version 1.6.2 should work, although newer versions are supported as well.

In order to teleoperate the robot we can use the launch file of the TurtleBot, as REEM is compatible with it. If you do not have the TurtleBot installed yet, just install it with the following command:

```
sudo apt-get install ros-groovy-turtlebot ros-groovy-turtlebot-apps
```

Then, just run the following command:

```
rosrun turtlebot_teleop turtlebot_teleop_key
```

Run the following command in order to relay the velocity commands to REEM:

```
rosrun topic_toolrelay turtlebot_teleop/cmd_vel cmd_vel
```

Use the keyboard to move the robot. At the same time you can run `rviz` to visualize the robot sensors, as shown in the next figure (in `cfg` there is a `reem.rviz` file to set up the `rviz` interface for REEM):

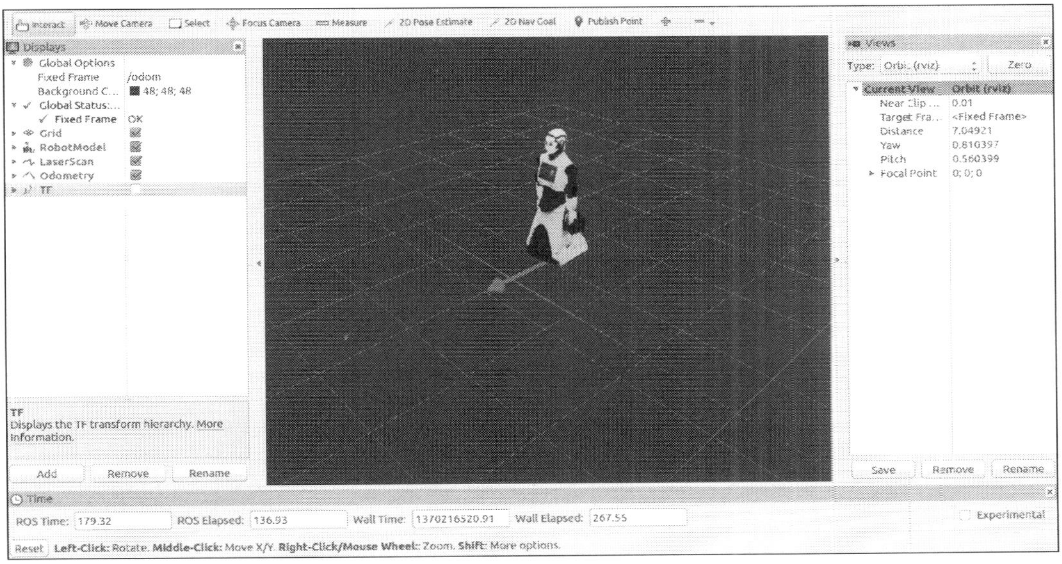

Similarly, apart from the URDF model, we can see all the TF and frames of REEM, as shown in the next figure:

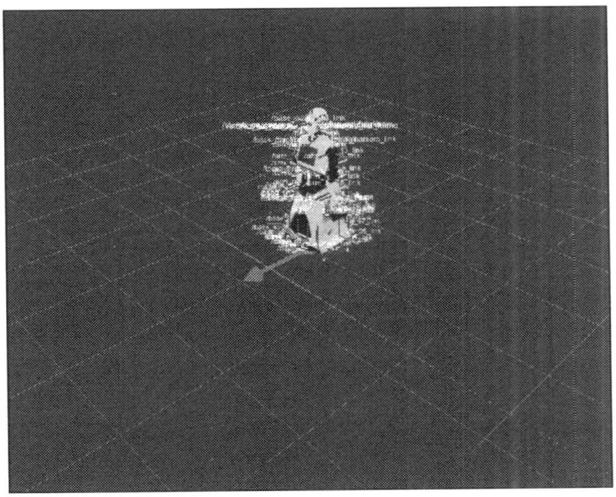

Finally, the following screenshot shows the robot after moving to create a map of the world simulated in Gazebo. At this point, in the screenshot we see the robot navigating in the map built, using the path planners available in ROS and the particles that estimate the robot localization within the map.

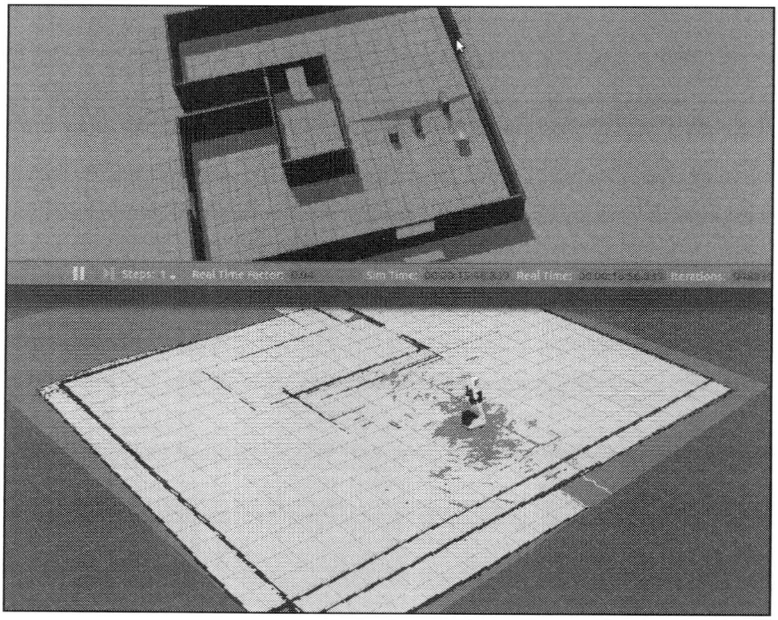

PR2 – the Willow Garage robot

PR2 is the mobile manipulation robot developed by Willow Garage, the developers of ROS. The robot is equipped with two arms and a head with stereo cameras for long and short distances, as well as a tilting laser.

Installing the PR2 simulator

We can install R2 using a Debian binary directly with the following command:

```
sudo apt-get install ros-fuerte-nasa-r2-simulator
```

Alternatively, we can install from sources by following these instructions.

To use the PR2 simulator, which uses Gazebo and the URDF model of the PR2 robot, just run the following:

```
sudo apt-get install ros-groovy-pr2-simulator
```

This will install several stacks and packages that allow you to move the robot and control its arms. This functionality is required by some demos not included in this book, for example, one that shows how to open doors with the PR2 robot.

Note that for the PR2 we also require ROS Groovy, although you can try other distributions. Even more important is to use a new version of Gazebo and check that it runs correctly, at least outside ROS, by running the following:

```
gazebo
```

What happens is that the newest versions of Gazebo are actually included in the newest distributions of ROS. So that is the easiest way to install it and to obtain the best integration among them.

Running PR2 in simulation

Here we will see how to run the simulator and see the PR2 model, as well as moving the base and the arms. We can simply run load PR2 in the empty world in Gazebo with:

```
roslaunch pr2_gazebo pr2_empty_world.launch
```

Downloading the example code

You can download the example code files for all Packt books you have purchased from your account at `http://www.packtpub.com`. If you purchased this book elsewhere, you can visit `http://www.packtpub.com/support` and register to have the files e-mailed directly to you. You can also download these code files from `https://github.com/AaronMR/Learning_ROS_for_Robotics_Programming`.

Similarly, we can launch the PR2 alone or in other worlds using the several launch files in the `launch` folder inside the `pr2_gazebo` package. For the empty world we will see the robot as shown in the following screenshot:

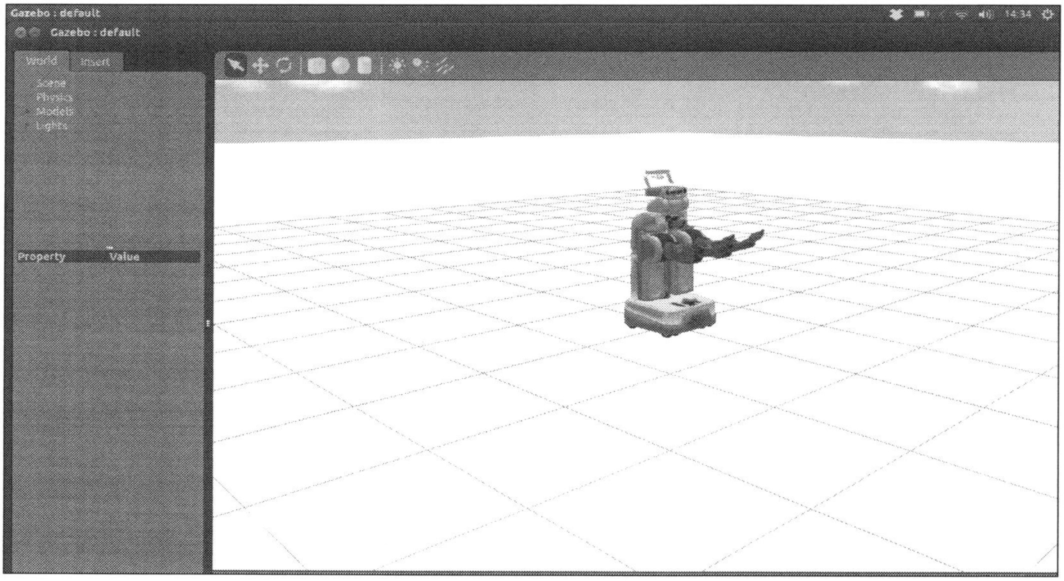

Note that Gazebo is already running some launch files that are required. It is worth mentioning that we can load the robot without the arms, which can reduce the system load if we are not going to use them. In order to run PR2 without arms run the following command:

roslaunch pr2 pr2_no_arms_empty_world.launch

This uses the launch file in the `chapter9_tutorials`, which has the following script (based on `pr2_empty_world.launch`):

```
<launch>
  <!-- start up empty world -->
  <arg name="gui" default="true"/>
  <arg name="throttled" default="false"/>
  <arg name="paused" default="true"/>

  <include file="$(find gazebo_worlds)/launch/empty_world_paused.
launch">
    <arg name="gui" value="$(arg gui)" />
    <arg name="throttled" value="$(arg throttled)" />
```

```
    <arg name="paused" value="$(arg paused)" />
  </include>

  <!-- start pr2 robot w/o arms -->
  <include file="$(find pr2_gazebo)/launch/pr2_no_arms.launch"/>
</launch>
```

We will see the following in Gazebo:

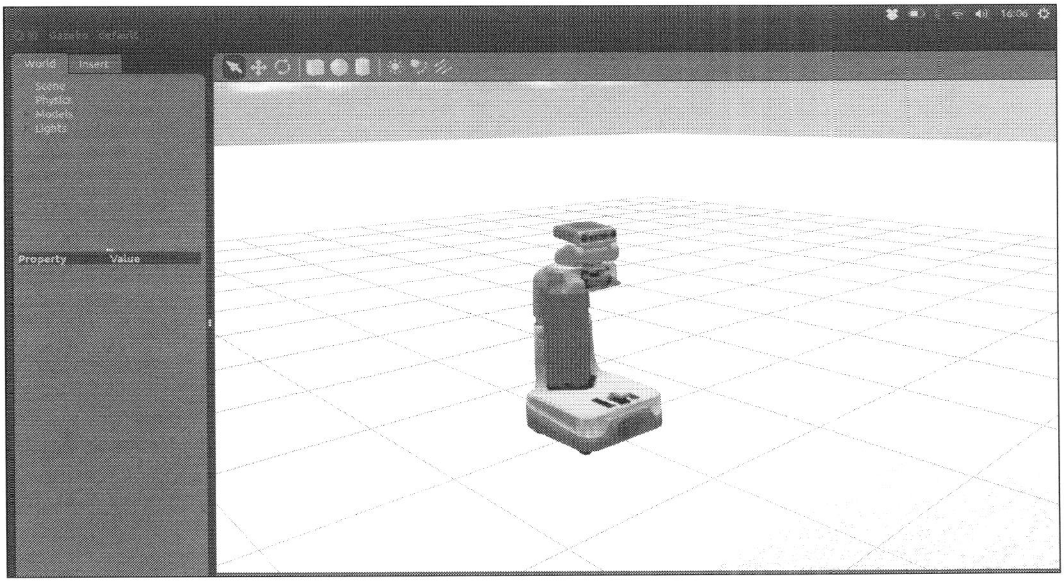

In order to move (teleoperate) the PR2, we will launch the simulation:

```
roslaunch pr2_gazebo pr2_empty_world.launch
```

We then teleoperate the robot with the keyboard or the joystick using one of the nodes provided with this Debian package:

```
sudo apt-get install ros-groovy-pr2-teleop-app
```

If you use the keyboard, run:

```
rosrun pr2_teleop teleop_pr2_keyboard
```

You also have a node for a joystick, and launch files that include each of these nodes.

We have to connect the `/cmd_vel` topic with the PR2 `/base_controller/command`, as follows:

```
rosrun topic_tools relay /cmd_vel /base_controller/command
```

We can check that the velocity commands are received by the robot by echoing the `Twist` messages:

```
rostopic ho /base_controller/command
```

The moving robot is shown in the following screenshot:

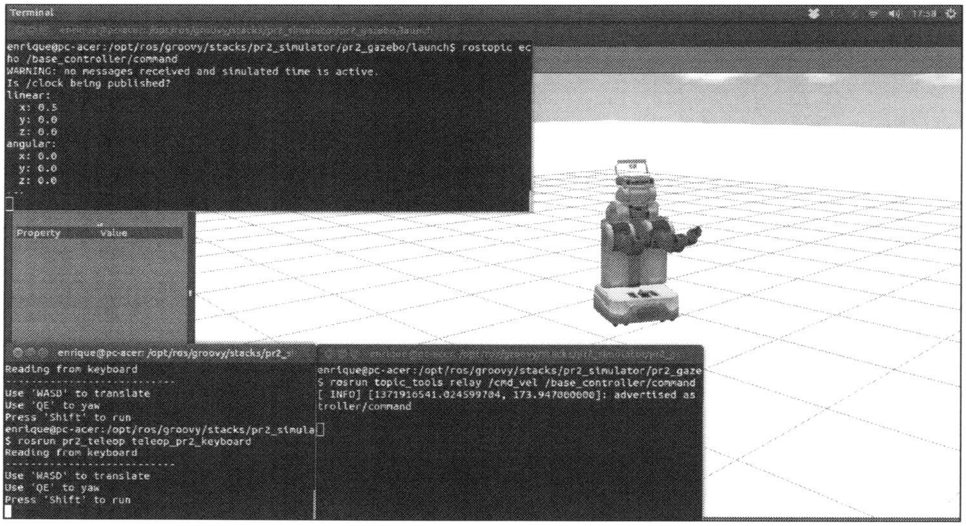

Finally, we can visualize the robot in `rviz`, using the `/odom_combined` frame, as shown in the following screenshot:

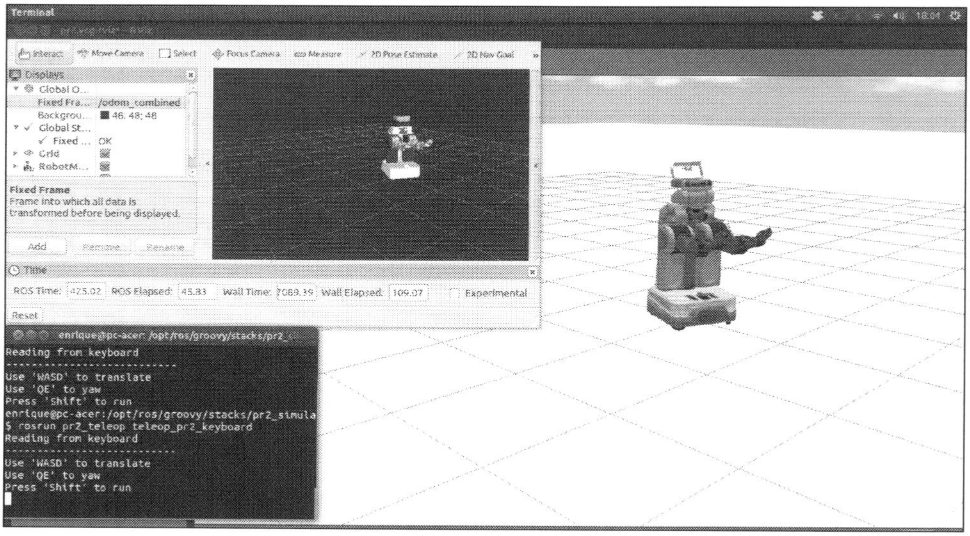

PR2 is perfectly integrated with the ROS navigation stack, so we can load a world in Gazebo and localize the robot using `amcl`, as well as mapping, while the robot moves on a given goal for the path planner.

Localization and mapping

Given a world in simulation, we can map it and localize the robot within it. There are several models and worlds available at the following link: `http://gazebosim.org/models/`.

We can reference them in the world file, and they will be automatically downloaded into our `.gazebo` home folder (at `$HOME`). Then we can also insert them manually. Alternatively we can design our own world in Gazebo, in Building Editor. We are going to use the simple room shown as follows:

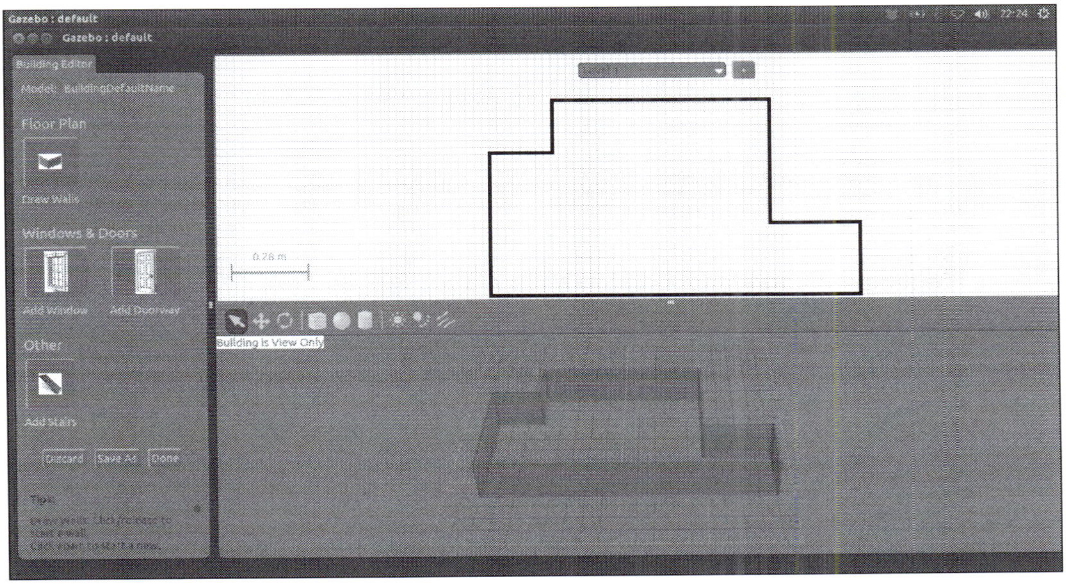

More complex worlds already exist and can be used, but it will take more time to map them completely. See the example of the Willow Garage world as follows, for instance, which is loaded in Gazebo with the following:

```
roslaunch pr2_gazebo pr2_wg_world.launch
```

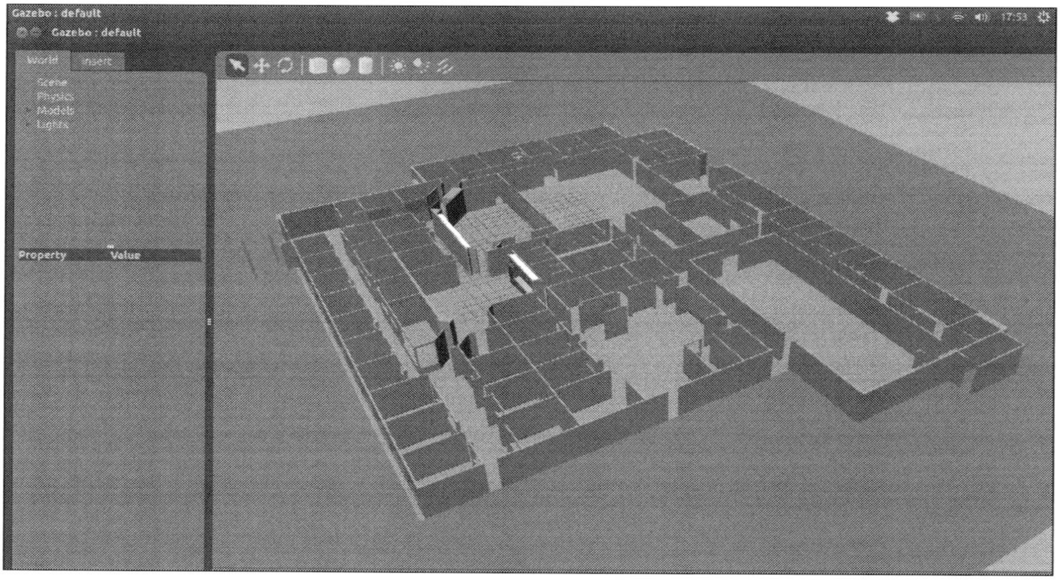

Our simple room world is saved as a `.sdf` file that we save as a world file for convenience; otherwise it will be a model that we could spawn in Gazebo. We provide a launch file to load the world:

```
roslaunch pr2 simple_world.launch
```

Also one to spawn the PR2 robot model into the world (note that you might see some rendering artifacts in some versions of Gazebo, but everything should work correctly):

```
roslaunch pr2 pr2_no_arms_simple_world.launch
```

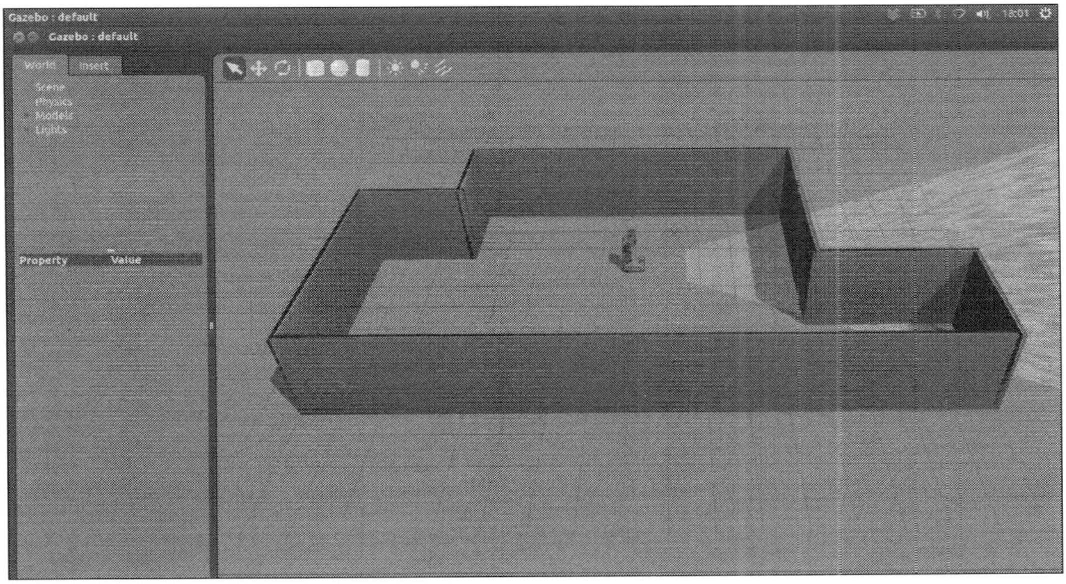

Now we need a map to navigate the environment. We are going to use the `gmapping` algorithm to do so. Run the following launch file, provided with the book:

```
roslaunch pr2 pr2_mapping.launch
```

It simply loads the previous world and the `gmapping` node using a launch file that comes with this package for the PR2 robot, with the appropriate tuning of the algorithm parameters.

Now, we must move the robot manually. Use the following launch files to do so, whether you use a joystick or the keyboard, respectively, choose one.

```
roslaunch pr2_teleop teleop_joystick.launch
roslaunch pr2_teleop teleop_keyboard.launch
```

You can see how the map is built in `rviz`; we run it from the launch file to start it with an adequate configuration to see the map while it is being built. Once the map is built and you are satisfied with the result, you can save the map with the following:

```
rosrun map_server_map_saver
```

This will create a `map.yaml` and a `map.pgm` file, which have the map characteristics and the map as an image in the same format of an **Occupancy Grid Map** (**OGM**). The files are created in the current folder, and then you can move them to the `maps` folder in the `pr2` folder of this chapter. There you will see two files that come with the book, which were created for your convenience. These files are used for the next part: the localization.

For the robot localization, and path planning we use the **Adaptive Monte Carlo Localization** (**AMCL**) and the `move_base` package, along with `costmap_2d`, which provides the costmap of the surrounding area that allows the robot to move without crashing into the obstacles in the environment.

You can run the following launch file which will automatically load the map built previously, as long as it is in the `maps` folder.

```
roslaunch pr2 pr2_navigation.launch
```

This runs the `map_server` node with the `maps/map.yaml` file, `amcl`, and the `move_base` nodes.

Running the demos of the PR2 simulator

Once we have a map and we are able to localize and move within the map, we can run some demos provided for the PR2. These demos comprise some tasks such as grasping and interacting with the environment.

You can open doors with the PR2, see how the robot plugs itself in to the electric supply, or just see it grasping objects from a table. Just apply all you have learned in this book and follow the instructions given in the following links:

- `http://ros.org/wiki/pr2_simulator/Tutorials/PR2OpenDoor/diamondback`
- `http://ros.org/wiki/pr2_simulator/Tutorials/PR2PluggingIn`
- `http://ros.org/wiki/pr2_simulator/Tutorials/SimpleGraspingDemo`

In some cases you will have to download the sources and compile them, but you can also install the Willow Garage PR2 applications:

```
sudo apt-get install ros-groovy-wg-pr2-apps
```

It includes some of the examples listed previously, among others.

Robonaut 2 – the dexterous humanoid of NASA

Robonaut 2, or R2 for short, is the second version of the dexterous humanoid robot developed by NASA. This robot is meant to help humans work in space. Its main feature is the ability to perform dexterous manipulation. The agency has built four robonauts at the time of writing this book. Here we present R2, which is supported in simulation inside ROS, using the software available online.

The R2 shown in the previous picture was launched to the International Space Station (ISS) as part of the STS-133 mission, and it is the first robot of its kind in space. It is important to note that R2 is deployed on a fixed pedestal inside the ISS, as we will see in the simulation environments in the sequel. However, other robonauts will be attached to a four-wheeled vehicle, or will have legs in the future.

Installing the Robonaut 2 from the sources

First of all, we must download the sources, which are in several repositories in bitbucket. We have two options. One consists of downloading each repository separately, as follows:

- `git clone https://bitbucket.org/nasa_ros_pkg/nasa_rosdep.git`
- `git clone https://bitbucket.org/nasa_ros_pkg/nasa_r2_common.git`
- `git clone https://bitbucket.org/nasa_ros_pkg/nasa_r2_simulator.git`

The other option uses `rosinstall`, which actually performs the three operations mentioned. For ROS Groovy we will simply run the following code:

```
Rosinstall . https://bitbucket.org/nasa_ros_pkg/misc/raw/master/nasa-ros-pkg.rosinstall
```

In any case, do not forget to update the stacks and packages databases:

```
rosstack profile && rospack profile
```

Once we have downloaded the R2 sources, we proceed to compile them:

```
roscd nasa_r2_simulator
```

You can use the `master` branch, but in case you are using ROS Groovy, you will need to use the `groovy` branch, so run the following:

```
git checkout groovy
```

Then you can build the code as follows:

```
rosmake
```

This will compile the R2 simulator, which depends on several packages from this and the other stacks (contained in each of the repositories downloaded).

Running Robonaut 2 in the ISS fixed pedestal

The most basic example launches the Gazebo simulator with the R2 in an empty environment.

```
roslaunch r2_gazebo r2_empty.launch
```

You will see the R2 in the Gazebo empty world, as shown in the following picture:

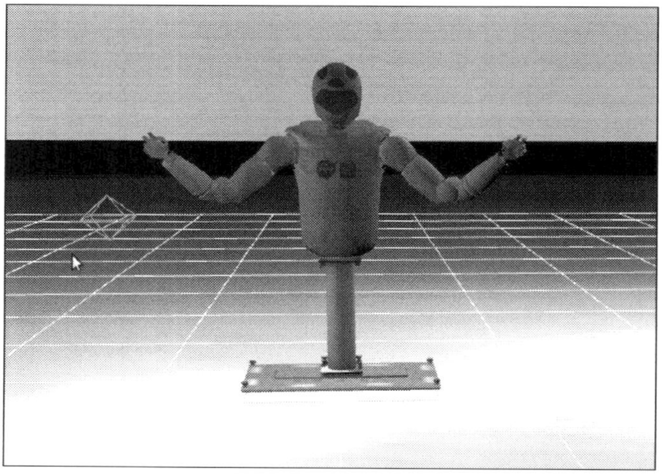

Controlling the Robonaut 2 arms

The first thing we must do is put the robot in the ready position

```
rosrun r2_controllers_gazebo r2_readypose.py
```

Then we can choose between the two arm control models, which are common for every robotic arm.

- **Joint mode**: We set the angle of each joint individually.
- **Cartesian mode**: We give the desired pose of a link, for example, the end effector, and the driver solves the inverse kinematics to obtain the appropriate joint state that allows the link to be at such a pose, as long as it is actually feasible.

By default, the robot starts in joint mode, but it is easy to change between modes. To enter the Cartesian mode, just run the following:

```
rosservice call /r2_controller/set_tip_name <arm_name> <link_name>
```

Here `arm_name` is either `left` or `right`, and `link_name` is any link in the respective arm chain. For example, the left middle finger is the link `left_middle_base`.

Similarly, to enter the joint mode, just run:

```
rosservice call /r2_controller/set_joint_mode <arm_name>
```

Then, when we are in each of the modes, we can control the arm pose using topics. In the case of the joint mode, we can publish `JointState` to either of these topics, which correspond to the left and right arm respectively.

```
/r2_controller/left_arm/joint_command
/r2_controller/right_arm/joint_command
```

The head and waist are controlled solely with `JointState`. On one side, the head is a pan-tilt device, which can be controlled with the topic `/r2_controller/neck/joint_command`; meanwhile, the waist is accessed through the topic `/r2_controller/waist/joint_command`.

Controlling the robot easily with interactive markers

The most flexible way to control the robot is using interactive markers within `rviz`. With the R2 simulator (Gazebo) running, open `rviz`:

```
rosrun rviz rviz
```

And then, open the interactive control teleoperator:

```
rosrun r2_teleop r2_interactive_control.py
```

You can use the `vcg` file provided with the book in `r2/config` to set up `rviz` appropriately for the Robonaut 2 and Interactive Markers. You can see two positions of the arms in the following screenshot:

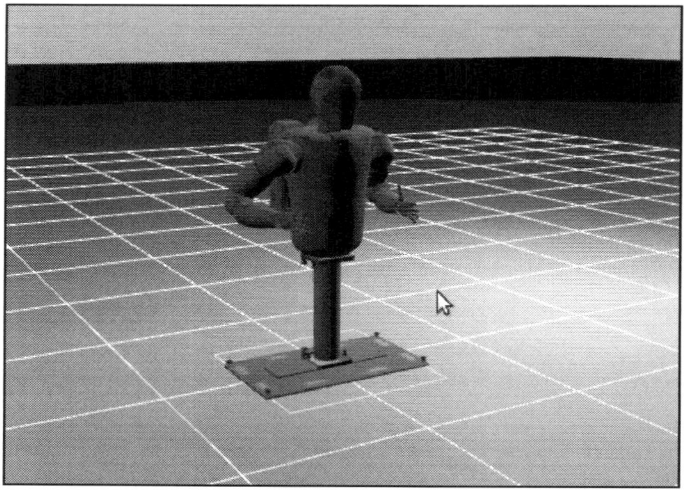

And with the interactive markers we have the following image:

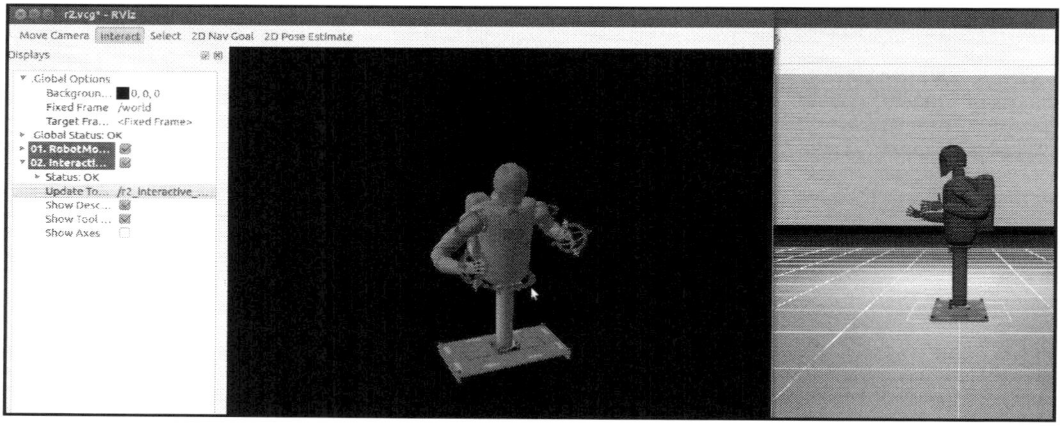

We observe the interactive markers that we can control directly in `rviz`, which allow us to give the desired position of the hands, and by inverse kinematics of the arm chain the joint's state is obtained so the arm reaches the desired pose.

Giving legs to Robonaut 2

In simulation we can see the legged version of R2, or simply the legs. The full body is launched with the following:

```
roslaunch r2_gazebo r2c_full_body.launch
```

If you prefer to work with the legs only, just run the following:

```
roslaunch r2_gazebo r2c_legs.launch
```

These legs are the R2 IVA climbing legs of a real Robonaut model.

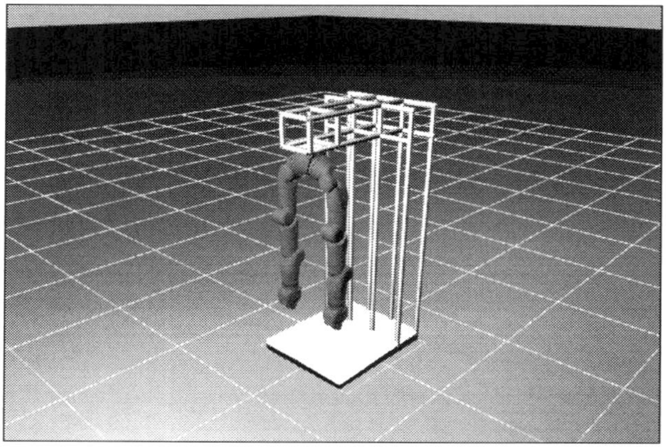

The Robonaut 2 comes with several Gazebo plugins for this purpose. They are automatically loaded/used by `gazebo`.

Loading the ISS environment

We can also load the ISS environment along with R2:

```
roslaunch r2_gazebo r2_ISS.launch
```

You will see the R2 on the pedestal inside of the ISS world loaded in Gazebo, as shown in the following screenshot:

This may consume many resources in your system, so you can load only the ISS Task Board with the following:

```
roslaunch r2_gazebo r2_taskboard.launch
```

In the following screenshot you can see the R2 with the Task Board only, which is actually the one we have in the ISS world, but without the rest of the elements of the ISS model:

Husky – the rover of Clearpath Robotics

The Husky A200 is an **Unmanned Ground Vehicle** (**UGV**) developed by Clearpath Robotics, one of the spin-off companies of Willow Garage. As we can see in the following picture, it is a rugged, outdoor vehicle, which is equipped with an Axis camera, GPS, and integrates the ROS framework:

Installing the Husky simulator

Here we focus on running Husky on simulation, so just run the following:

```
sudo apt-get install ros-groovy-husky-simulator
```

Alternatively, you can clone the repository in order to have the latest version:

```
git clone git://github.com/clearpathrobotics/husky_simulator.git
```

Running Husky on simulation

We load an empty map in Gazebo:

```
roslaunch husky_gazebo map_empty.launch
```

Now just load Husky with the following:

```
roslaunch husky_gazebo husky_base.launch
```

This can also be done manually within Gazebo. You will see the robot as shown in the following screenshot:

Then, we can use the PR2 tele-operation node to move it, so just run the following:

```
rosrun pr2_teleop teleop_pr2_keyboard
```

Note that the Husky robot is not holonomic, so it cannot move laterally. Therefore, if you press the keys *A* or *D*, it will not move. The next screenshot shows Husky after moving it.

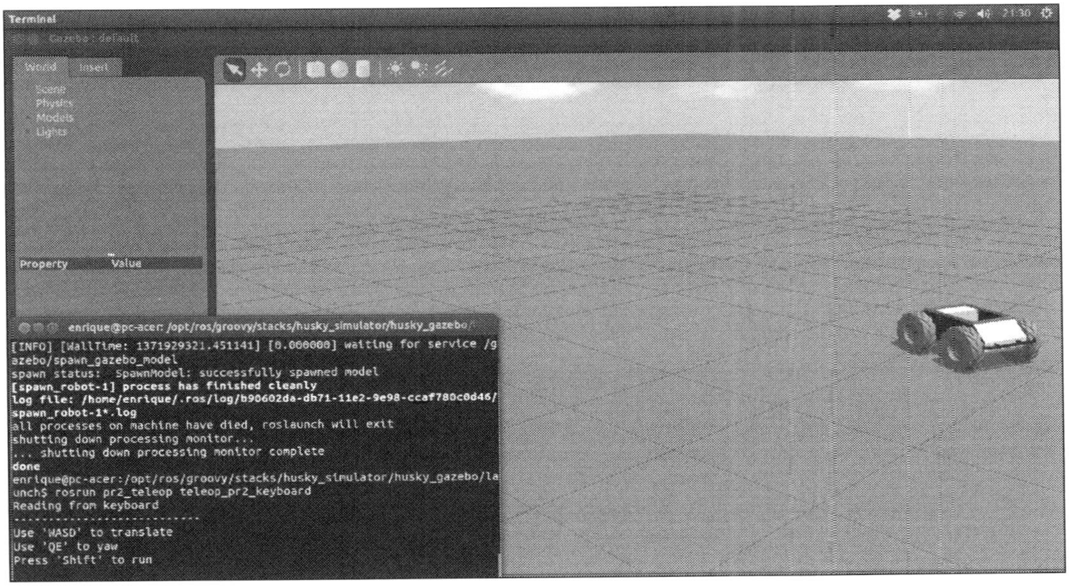

Now, as Husky integrates with the navigation stack, you can localize it, map a simulated world, and navigate through it. You only have to apply what you have learned in this book.

TurtleBot – the low-cost mobile robot

TurtleBot integrates in several simple and low-cost state-of-the-art hardware devices such as Microsoft's Kinect sensor, Yujin Robot's Kobuki, and iRobot's Create. It is a moving platform intended for indoor environments and with 3D perception capabilities by means of the Kinect. It is also easy to include more devices or even use with our own laptop to test our software, as it has several platforms at different heights, as the next picture shows:

Installing the TurtleBot simulation

The simulator of the TurtleBot does not come with the ROS repositories, so clone the `turtlebot_simulator` repository:

```
git clone https://github.com/turtlebot/turtlebot_simulator.git
```

Use the `master` branch or move to the correct one for your ROS distro. For example, for ROS Groovy do the following:

```
git checkout groovy
```

Now, build the sources:

```
rosmake
```

Running TurtleBot on simulation

To run TurtleBot in the empty world run the following code:

```
roslaunch turtlebot_gazebo turtlebot_empty_world.launch
```

As you see, the typical standard is followed since the robot can be spawned in a `gazebo` world using the launch files in the <robot_name>_gazebo package. As with the other robots, now you can perform mapping, localization, and navigation tasks with it, as well as inspecting virtual worlds with the sensors of the robot.

Summary

In this chapter we have seen several real robots from international companies and how to work with them in simulation. Here you can use all the background learnt in all the previous chapters of this book. Now you can play with almost the same tools and robots that the most important robotics institutions and companies develop.

We have shown several humanoid robots, such as REEM, PR2, and Robonaut. These robots allow you to grasp objects, since they incorporate arms in the torso. Similarly, they can move the head and you can focus on particular zones of the environment. All three include wheeled-driven robots, holonomic or not, and the developers of REEM and Robonaut also have biped robots. Indeed, REEM-C and R2 are legged robots that can walk.

Later we have shown other robotic platforms that only include a base with wheels and probably some supporting structures to attach sensors or an external laptop. This is the case of Husky and TurtleBot. In fact, TurtleBot is one of the cheapest robots of its kind. Here you have learned how to run it in simulation, but you might also buy one and do SLAM, localization, path planning, and all the exciting things you have learnt in this book. You do not need anything else.

Now is your turn to do ground-breaking stuff with your own robots or with the most popular ones, which are already integrated in ROS, as you have seen here. Happy robot programming!

Index

Symbols

A

B

C

D

E

F

G

H

S

T

U

V

W

X

Thank you for buying

Learning ROS for Robotics Programming

About Packt Publishing

Packt, pronounced 'packed', published its first book "*Mastering phpMyAdmin for Effective MySQL Management*" in April 2004 and subsequently continued to specialize in publishing highly focused books on specific technologies and solutions.

Our books and publications share the experiences of your fellow IT professionals in adapting and customizing today's systems, applications, and frameworks. Our solution based books give you the knowledge and power to customize the software and technologies you're using to get the job done. Packt books are more specific and less general than the IT books you have seen in the past. Our unique business model allows us to bring you more focused information, giving you more of what you need to know, and less of what you don't.

Packt is a modern, yet unique publishing company, which focuses on producing quality, cutting-edge books for communities of developers, administrators, and newbies alike. For more information, please visit our website: `www.packtpub.com`.

About Packt Open Source

In 2010, Packt launched two new brands, Packt Open Source and Packt Enterprise, in order to continue its focus on specialization. This book is part of the Packt Open Source brand, home to books published on software built around Open Source licences, and offering information to anybody from advanced developers to budding web designers. The Open Source brand also runs Packt's Open Source Royalty Scheme, by which Packt gives a royalty to each Open Source project about whose software a book is sold.

Writing for Packt

We welcome all inquiries from people who are interested in authoring. Book proposals should be sent to `author@packtpub.com`. If your book idea is still at an early stage and you would like to discuss it first before writing a formal book proposal, contact us; one of our commissioning editors will get in touch with you.

We're not just looking for published authors; if you have strong technical skills but no writing experience, our experienced editors can help you develop a writing career, or simply get some additional reward for your expertise.

[PACKT] open source
PUBLISHING
community experience distilled

C Programming for Arduino

ISBN: 978-1-849517-58-4 Paperback: 512 pages

Learn how to program and use Arduino boards with a series of engaging examples, illustrating each core concept

1. Use Arduino boards in your own electronic hardware and software projects
2. Sense the world by using several sensory components with your Arduino boards
3. Create tangible and reactive interfaces with your computer
4. Discover a world of creative wiring and coding fun!

TortoiseSVN 1.7 Beginner's Guide

ISBN: 978-1-849513-44-9 Paperback: 260 pages

Perform version control in the easiest way with the best SVN client – TortoiseSVN

1. Master version control techniques with TortoiseSVN without the need for boring theory
2. Revolves around a real-world example based on a software company
3. The first and the only book that focuses on version control with TortoiseSVN
4. Reviewed by Stefan Kung, lead developer for the TortoiseSVN project

Please check **www.PacktPub.com** for information on our titles

31059182R00187

Made in the USA
Lexington, KY
28 March 2014